IET POWER AND ENERGY SERIES 46

Series Editors: Professor A.T. Johns
Professor D.F. Warne

Distribution Switchgear

Other volumes in this series:

Volume 1	**Power circuit breaker theory and design** C.H. Flurscheim (Editor)
Volume 4	**Industrial microwave heating** A.C. Metaxas and R.J. Meredith
Volume 7	**Insulators for high voltages** J.S.T. Looms
Volume 8	**Variable frequency AC motor drive systems** D. Finney
Volume 10	**SF6 switchgear** H.M. Ryan and G.R. Jones
Volume 11	**Conduction and induction heating** E.J. Davies
Volume 13	**Statistical techniques for high voltage engineering** W. Hauschild and W. Mosch
Volume 14	**Uninterruptable power supplies** J. Platts and J.D. St Aubyn (Editors)
Volume 15	**Digital protection for power systems** A.T. Johns and S.K. Salman
Volume 16	**Electricity economics and planning** T.W. Berrie
Volume 18	**Vacuum switchgear** A. Greenwood
Volume 19	**Electrical safety: a guide to causes and prevention of hazards** J. Maxwell Adams
Volume 21	**Electricity distribution network design, 2nd edition** E. Lakervi and E.J. Holmes
Volume 22	**Artificial intelligence techniques in power systems** K. Warwick, A.O. Ekwue and R. Aggarwal (Editors)
Volume 24	**Power system commissioning and maintenance practice** K. Harker
Volume 25	**Engineers' handbook of industrial microwave heating** R.J. Meredith
Volume 26	**Small electric motors** H. Moczala *et al.*
Volume 27	**AC-DC power system analysis** J. Arrill and B.C. Smith
Volume 29	**High voltage direct current transmission, 2nd edition** J. Arrillaga
Volume 30	**Flexible AC Transmission Systems (FACTS)** Y-H. Song (Editor)
Volume 31	**Embedded generation** N. Jenkins *et al.*
Volume 32	**High voltage engineering and testing, 2nd edition** H.M. Ryan (Editor)
Volume 33	**Overvoltage protection of low-voltage systems, revised edition** P. Hasse
Volume 34	**The lightning flash** V. Cooray
Volume 35	**Control techniques drives and controls handbook** W. Drury (Editor)
Volume 36	**Voltage quality in electrical power systems** J. Schlabbach *et al.*
Volume 37	**Electrical steels for rotating machines** P. Beckley
Volume 38	**The electric car: development and future of battery, hybrid and fuel-cell cars** M. Westbrook
Volume 39	**Power systems electromagnetic transients simulation** J. Arrillaga and N. Watson
Volume 40	**Advances in high voltage engineering** M. Haddad and D. Warne
Volume 41	**Electrical operation of electrostatic precipitators** K. Parker
Volume 43	**Thermal power plant simulation and control** D. Flynn
Volume 44	**Economic evaluation of projects in the electricity supply industry** H. Khatib
Volume 45	**Propulsion systems for hybrid vehicles** J. Miller
Volume 46	**Distribution switchgear** S. Stewart
Volume 47	**Protection of electricity distribution networks, 2nd edition** J. Gers and E. Holmes
Volume 48	**Wood pole overhead lines** B. Wareing
Volume 49	**Electric fuses, 3rd edition** A. Wright and G. Newbery
Volume 50	**Wind power integration: connection and system operational aspects** B. Fox *et al.*
Volume 51	**Short circuit currents** J. Schlabbach
Volume 52	**Nuclear power** J. Wood
Volume 53	**Condition assessment of high voltage insulation in power system equipment** R.E. James and Q. Su
Volume 55	**Local energy: distributed generation of heat and power** J. Wood
Volume 56	**Condition monitoring of rotating electrical machines** P. Tavner, L. Ran, J. Penman and H. Sedding
Volume 905	**Power system protection, 4 volumes**

Distribution Switchgear

Stan Stewart

The Institution of Engineering and Technology

Published by The Institution of Engineering and Technology, London, United Kingdom

First published 2004

The Institution of Engineering and Technology
Michael Faraday House
Six Hills Way, Stevenage
Herts, SG1 2AY, United Kingdom

www.theiet.org

British Library Cataloguing in Publication Data
Stewart, Stan
Distribution switchgear
1. Electric switchgear
I. Title
621.3'17

ISBN (10 digit) 0 85296 107 3
ISBN (13 digit) 978-0-85296-107-0

Typeset in India by Newgen Imaging Systems (P) Ltd, Chennai
First printed in the UK by MPG Books Ltd, Bodmin, Cornwall
Reprinted in the UK by Lightning Source UK Ltd, Milton Keynes

This book is dedicated to those engineers who gave me help, guidance, encouragement and sound advice in my formative years. It is also dedicated to my grandchildren Hattie, Hannah, Robert and Tom, in the hope that they will be as happy and fortunate as I was, in their choice of careers.

Contents

Foreword

Each day, during which something new is learned, should be counted as a good day, and every engineer worth his salt, should keep a notebook to record the things learned for future reference. These notebooks should give references to sources of information, and most importantly, contain careful notes of those things not covered by textbooks. It was, therefore, with some humility that I accepted the invitation from the IEE to write this book, as I was very conscious of all the brilliant, talented engineers and scientists who were around in my formative years, who are now gone and never published their personal engineering notes.

As a result of nearly fifty years of personal involvement in the switchgear industry, most of which being in switchgear design and development, my personal notes now run to several volumes. So when I was asked if I would prepare this book I took the opportunity to include a number of interesting and directly useful items from those notes that I hope practising engineers will find useful, and which otherwise, in time may have been eventually lost.

You will find that certain chapters within the book are dedicated to what can best be described as components of distribution switchgear. These are Chapters 8 and 12 which deal with operating mechanisms and high-voltage fuse-links, respectively.

I did this for the simple reason that they form an essential component within distribution switchgear and fuse-switchgear and therefore merit a chapter in their own right. However, in the case of high-voltage fuses, I direct the reader to the IEE Power Book on the subject in Chapter 17.

Apart from the basic knowledge building chapters, such as interruption techniques and fault level calculations, the book contains a number of interesting and useful subjects that had to be learned the hard way. For example,

- Under what conditions, can well designed electrical insulation fail in service with normal levels of electrical stress?
- For a given short-circuit current, how do you know what minimum contact loads you need to prevent contact burning?
- How do you set about designing a magnetic actuator operating mechanism?
- How does a rotating arc SF_6 interrupter work?
- How do you calculate the cost of ownership of switchgear?

As the answers to these questions, and many more, will be found within this book, I hope that you will find them useful and a source of reference. Perhaps even adding them to your own library of notes.

Stan Stewart,
Cheshire, England
June, 2003

Acknowledgements

In particular, I would like to thank Tony Turnbull of ALSTOM T&D Distribution Switchgear Ltd. for proof reading the draft and his very helpful suggestions. I would also like to thank the following people who responded to requests for information and illustrations:

Kristine Kucera of Cooper Power Systems, USA.
Tony Headley of the British Short Circuit Testing Station, UK.
Patty Kozlow of Joslyn Hi-Voltage, USA.
Martijn Venema of the KEMA Laboratories, Netherlands.
Saviour Zammit of Medelec Switchgear, Malta.
Claire Shore of Schneider Electric, UK.
Michael Jackson of ALSTOM T&D, South Africa.
Anne Busson of Lucy Switchgear, UK.
Phil Rosen and Gordon Newbery of Cooper Bussmann, Cooper UK Ltd.
Paul Miller, M&B System Sales Ltd UK.

Chapter 1
Basics and general principles

1.1 Why do we have switchgear?

A fundamental question is the type of question that children specialise in asking, and, in order to respond correctly and fully, you have to give the subject more thought than would otherwise be the case. My grandchildren specialise in these sorts of thought provoking questions and, as I am sure that all readers are familiar with the function of gooseberry bushes, I shall confine myself to try to address the fundamental question 'Why do we have switchgear?'.

Certain electrical distribution customer's senior engineers used to go out of their way to say that switchgear was a necessary evil. It cost money to buy, install and maintain and that it did not earn any revenue. This is clearly an oversimplification as the end user only buys electrical power, so anything that makes that possible must contribute to that end. These customer's engineers did, however, concede that switchgear was necessary to isolate equipment that became faulty, and they could allow the system to be split into sections to allow quick restoration of power supplies. While electricity distribution systems are relatively passive, for example, the situation in a factory, particularly one using manufacturing processes, or in a generating station can be active, the switchgear takes a critical part in controlling what is taking place.

So, switchgear is needed

(a) to isolate faulty equipment;
(b) to divide large networks into sections for repair purposes;
(c) to reconfigure networks in order to restore power supplies; and
(d) to control other equipment.

1.2 What is the difference between a circuit breaker and a switch?

All switchgear must be capable of either closing or opening an electrical circuit. This is defined in standards as: 'A general term covering switching devices and their combination with associated control, measuring, protection and regulating equipment'.

The question as to what the difference is between a circuit breaker and a switch can best be answered by first of all stating what they have in common.

(a) They both can carry and interrupt their rated normal current safely.
(b) They both can safely close their contacts onto a fault and carry that fault for a rated specified time.
(c) They both can safely withstand their rated power–frequency system voltage and rated lightning impulse voltage across their contacts when in the open position.

The difference between a circuit breaker and a switch is that a circuit breaker can detect and interrupt a short-circuit fault current, whereas a switch can do neither.

In addition to circuit breakers and switches, switchgear also includes the following.

1.2.1 Disconnectors

These are mechanical switches that, by definition, must be able to to carry a defined rated normal and short-circuit current, and in the open position must provide a defined level of insulation between their contacts. This will usually be an impulse voltage withstand level.

1.2.2 Earth switches

These are mechanical switches capable of carrying a rated short-circuit current. Unless they are off-load devices, they will also have the ability to make onto a rated peak short-circuit current and carry that current safely for a specified time.

1.2.3 Fuses

Fuses must be capable of carrying a defined load current without deterioration and be able to interrupt a defined short-circuit current. They may, or may not, include a mechanical tripping device such as a chemically propelled, or spring driven striker pin in order to trip its associated switch or indicate that it has operated.

Most switchgear items can exist in combinations. It is quite common to find items such as a switch-fuse or a switch disconnector. These combinations meet the individual technical requirements of the active elements within the combination.

1.3 Components of switchgear

Inherent within switchgear in the open position is the need for one side of the gap to be insulated from the other, and both sides to be always insulated from earth. If we put fuses to one side, we find that switchgear is usually in the form of a three-phase device. A simple single-phase diagram (Figure 1.1) illustrates the basic component functions within switchgear. It will be seen in Figure 1.1 that the basic

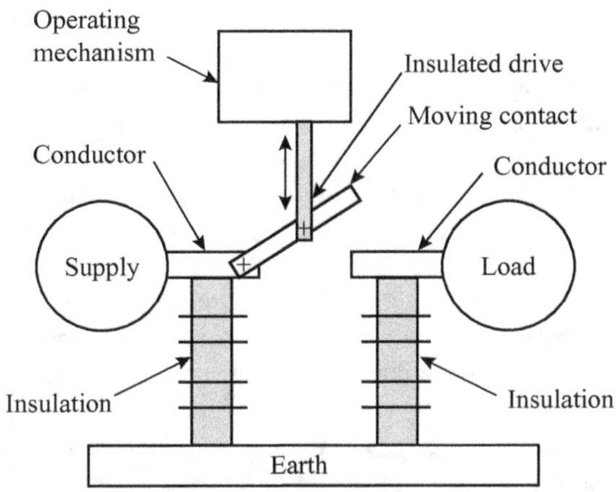

Figure 1.1 Basic components within switchgear

components within switchgear are:

(a) supply-side and load-side conductors;
(b) insulation from earth to support the conductors;
(c) a moving contact arranged to be able to join, or separate, the two conductors;
(d) a driving mechanism and its associated drive linkage to the moving contact.

The conductors (a) are required to carry electrical current and will, therefore, generate heat due to their internal resistance, and will also be subjected to mechanical forces due to the electromagnetic effects of electrical current. This latter subject is dealt with in Chapter 5.

The insulators (b) are required to provide electrical insulation to earth to withstand both the system voltage and any transient voltages which may be impressed upon the switchgear. It should be remembered that the insulation within switchgear is constantly electrically stressed throughout the life of the equipment. The insulation also has to withstand mechanical forces that may be transmitted to the insulation from the conductors.

The moving contact (c), like the conductors, will also be subjected to heat and mechanical forces. It can be expected that the potential for heat generation will be greater in the moving contact as it will have a transfer contact at its hinge, and spring-loaded contacts at its separable end. It is the function of the moving contact to provide an insulating gap when in the open position.

It is the function of the operating mechanism (c) to drive the moving contact between the open and closed positions, and to withstand the electromagnetic 'blow-off' forces that may be generated when a short circuit occurs. Operating mechanisms come in many different forms, but all of them are obliged to provide the power for operation independent of the rate at which the external power is supplied. This is to ensure that the contact speed during operation is constant. It will be appreciated that this is particularly important for manually charged operating mechanisms. Operating mechanisms are dealt with in Chapter 8.

Not shown on the diagram, but of great importance, is the means of extinguishing the arc that will always form when the moving contact is separated from its associated fixed contact while carrying current. It will be appreciated that when the contacts are in the closed position, the interrupting zone of the switchgear acts as a conductor, and in order to interrupt the flow of current, this conductor must change its condition to that of an electrical insulator. This change has to take place in the shortest possible time in order to minimise the effects of arcing. Depending upon the type of interrupting technology used, these effects may result in heat, gas and pressure generation as well as contact melting and erosion. Interrupting techniques are dealt with in Chapter 2.

In addition, and also not shown, are the earthing facilities. The functionality of a circuit breaker should include a means of earthing in order to allow safe working conditions on the unit's associated cable and/or busbars.

In practice, the architecture of switchgear will, to some degree, be dictated by the functionality that its application will demand. Figure 1.2 shows how the essential components are arranged in a typical horizontally isolated indoor circuit breaker. An actual embodiment of such an arrangement is shown in Figure 1.3.

A comparison of Figures 1.2 and 1.3 will allow the physical embodiment of the components to be identified. However, it will be noticed that the secondary wiring

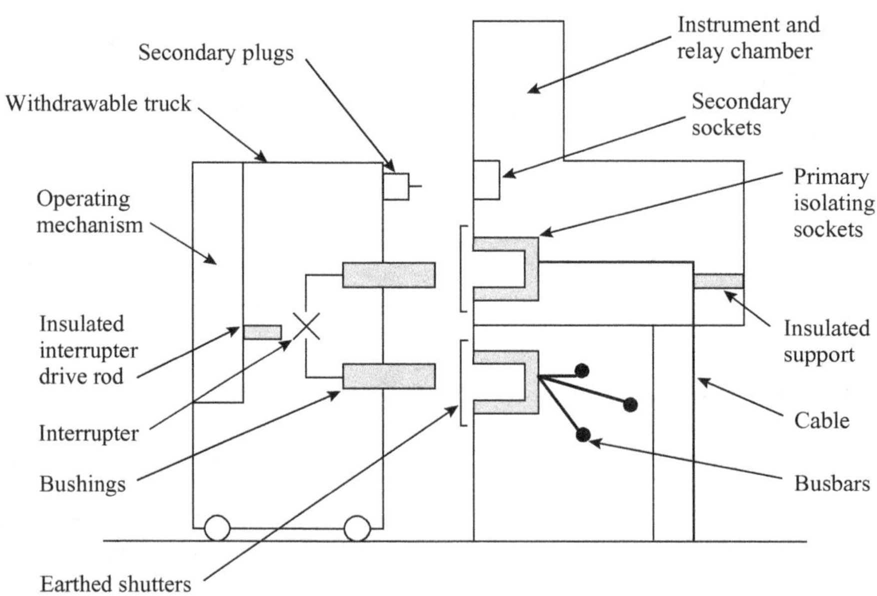

Figure 1.2 An arrangement of components in a horizontally isolated circuit breaker

Figure 1.3 Indoor horizontally isolated vacuum circuit breaker Type SVB5 (courtesy of ALSTOM South Africa)

connections shown in Figure 1.3 are via a cable connection, rather than a plug and socket attached to the moving and fixed portions, respectively. It will be appreciated that, with horizontally isolated switchgear, the earthing facilities will require either a separate earthing switch, as shown in Chapter 5 (Figure 5.6), or a means of raising and lowering of the circuit breaker element within its truck to facilitate transfer earthing using the circuit breaker itself. The type SVB5 shown in Figure 1.3 is believed to be unique in that, with the circuit breaker isolated, the connectors of the upper circuit breaker primary isolating contacts can be angled upwards so that when the circuit breaker is re-inserted, these contacts engage with a set of earthing contacts.

An alternative to the horizontally isolated circuit breaker is the vertically isolated type. This was extensively used within the United Kingdom and certain other markets at one stage, but is now less popular than the horizontally isolated type. The advantage of the vertically isolated circuit breaker is that, via a transfer position, the circuit breaker can be used for circuit and busbar earthing without the complication of separate earthing switches. A diagram of the essential components within a vertically isolated circuit breaker is shown in Figure 1.4.

Figure 1.4 shows the circuit breaker truck engaged in the normal service position. In order to provide safe working conditions on the cable, the circuit breaker would be opened and then lowered to disengage the primary isolating contacts. The circuit breaker truck would then be moved to its rear position, raised and then closed to earth the cable. A similar procedure would also be followed for earthing the busbars. The

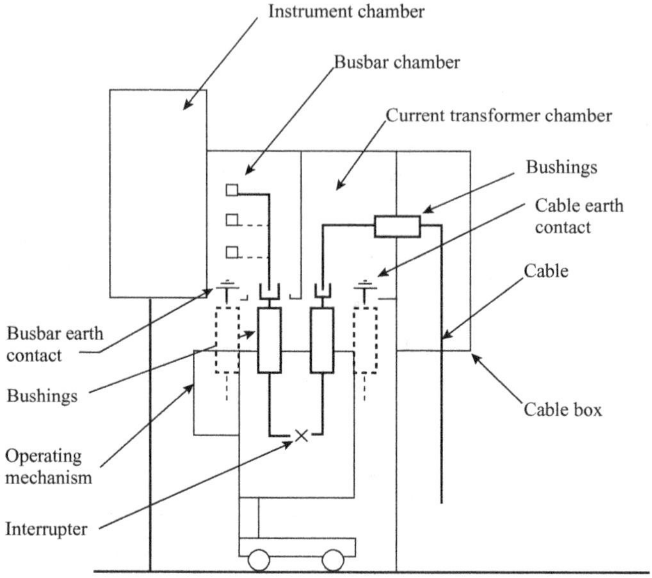

Figure 1.4 Components within a vertically isolated circuit breaker

Figure 1.5 A Type VMX switchboard of vertically isolated indoor circuit breakers (courtesy of ALSTOM T&D)

(*a*)

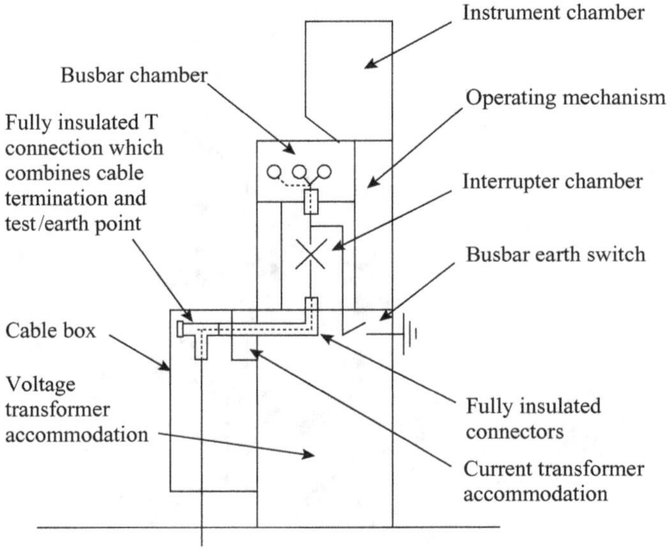

(*b*)

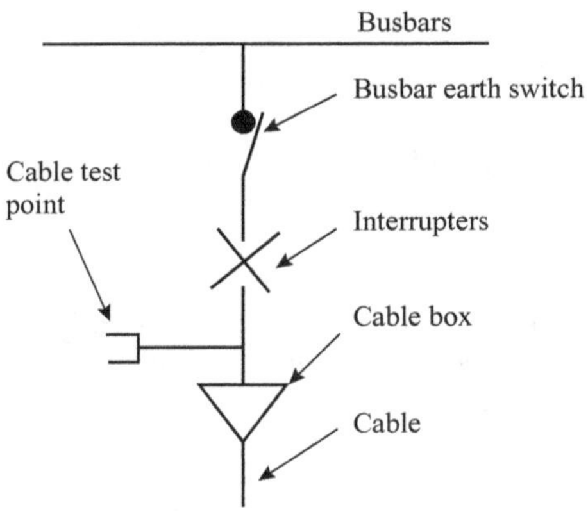

Figure 1.6 *(a) Components of a fixed circuit breaker (b) Line diagram of a fixed circuit breaker*

alternative positions are indicated in Figure 1.4. The photograph in Figure 1.5 shows a typical switchboard of vertically isolated switchgear. It will be seen in Figure 1.5 that two of the circuit breaker trucks are in the isolated (lower) position.

In recent years, there has been a tendency towards the use of fixed-type circuit breakers, particularly for secondary distribution. Fixed-type indoor circuit breakers have the attraction of offering the potential for a lower cost alternative arrangement of components. This trend has come about because it was recognised that modern circuit breakers have a very high reliability and it was argued that isolation of the

Figure 1.7 The 'Genie' fixed-type circuit breaker (courtesy of Groupe Schneider)

circuit breaker element is an unnecessary cost and complication in its construction. A counter-argument is that a fixed circuit breaker offers lower flexibility than an isolatable type. By this, it is meant that circuit breakers cannot be exchanged readily for maintenance, which could be important in critical locations such as within a process industry. Typical functional elements within a fixed circuit breaker type are shown in Figures 1.6(a) and (b).

A photograph of a compact fixed circuit breaker is shown in Figure 1.7.

Chapter 2
Interruption techniques

2.1 Arc interruption

Interruption of an alternating current arc, subtended between parted electrical contacts, will take place if the means for electrical re-ignition is removed. The gap between the contacts has to change from being an electrical conductor to being an electrical insulator at, ideally, a natural current zero.

There are a number of theories relating to the interruption of electrical current, and most of these are based upon the original theories of Cassie [1] or Slepian [2].

Cassie says:

> *If the energy lost from the arc column at current zero exceeds the energy input from the external electrical circuit, the electrical current will cease to flow.*

Slepian says:

> *If, after current zero, the dielectric strength of the contact gap increases at a greater rate than the transient voltage, then the circuit breaker will clear.*

Slepian's theory is illustrated in Figure 2.1.

A successful interruption is shown in Figure 2.1(a) where the rate of increase of dielectric recovery exceeds the rate of increase of the transient recovery voltage stress. Figure 2.1(b) shows a failure to clear as re-ignition occurs at a point where the impressed voltage exceeds the dielectric strength of the gap.

2.2 Interruption mediums and techniques

If we put to one side fault current interruption using high-voltage fuses, interrupting mediums used in medium voltage distribution switchgear today are oil, vacuum and SF_6 gas. There is a small percentage of units based upon hard gas, where the arc is forced into contact with materials that generate a gas to work on the arc and air break technology based upon cold cathode or insulated metal plates. However, techniques such as these are now very rare and will not be considered here. Oil interruption

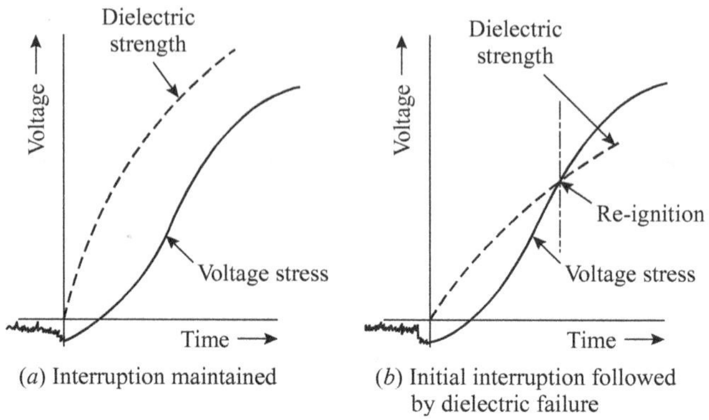

Figure 2.1 Slepian's theory of interruption and re-ignition

technology is no longer used for new primary switchgear applications, but, although in declining numbers, it is still used extensively within secondary switchgear. As the total population of circuit breakers is currently still dominated by oil interrupting types, it is important that the mechanism for arc interruption in oil is understood.

2.3 Oil switchgear

Until the 1970s, there was no real alternative to using oil filled switchgear for distribution applications. The origins of using oil as an electrical switching medium are unclear. It can only be assumed that the high dielectric strength of oil encouraged a pioneer to separate electrical contacts under oil. This was very successful and the assumption was that the oil quenched the arc. As a result, oil was widely taken up and used for many decades before the true reason for its effectiveness as an interrupting medium was discovered.

As electrical systems grew in size and power, they started to fail spectacularly. This resulted in ground breaking research being carried out by the ERA, which in the late 1920s at the Carville Power Station in the UK, determined what actually took place during oil interruption [3].

The experimental apparatus comprised a fixed and moving contact, submerged in oil, with a series of evacuated and sealed glass phials, arranged so that the neck of each phial was broken by the moving contact as it opened, allowing the phials to collect whatever gases were present. Analysis of the contents of the phials showed that the gases were predominantly hydrogen and acetylene. It was deduced that the effectiveness of oil circuit breakers was due to the presence of hydrogen, which, because of its low atomic weight, was capable of travelling at very high velocities and, therefore, provided a means for the rapid extraction of heat from the arc column. The oil circuit breaker was, in reality, a gas circuit breaker. This discovery helped engineers to understand that the mechanism of arc interruption involves rapid heat removal from the arc channel.

It is both interesting and alarming to note that the investigators carried on with their experiments by building and testing a hydrogen-filled circuit breaker, which confirmed their belief that it was the gas which enabled interruption and not, as originally thought, the oil. Fortunately, the danger of the hydrogen circuit breaker was fully realised at that time and there never was an intention to introduce a commercial version.

All oil circuit breakers are fitted with gas vent pipes. These are intended to vent the hydrogen produced during arc interruption, outside of the switchgear, or the substation, as there would be a danger of an internal explosion if the gas was ignited by, for example, a small arc in the circuit breaker auxiliary contacts. It is, therefore, very important that all gaskets and seals in the gas vent system are maintained in good condition. However, if the gas seals allow gas to enter the circuit breaker structure, it will take a finite time to disperse. A concentration of 5 per cent hydrogen or greater is required for the gas to ignite. Consider a circuit breaker that had cleared a fault and, as a result of ineffective gaskets, had partially been filled to 40 per cent with hydrogen gas.

Figure 2.2 shows that this circuit breaker would be liable to experience an internal explosion if a spark was produced when it was called upon to close within 8.5 min of the initial clearance.

Whenever visiting a substation having oil switchgear, it is good practice to examine the vent outlets of any circuit breakers, especially those set for autoreclose operations, as the presence of a small volume of oil, which would have been exhausted with the gas, would indicate that the circuit breaker has been working, and may need attention.

Oil circuit breakers were originally of the 'plain break' type. In this type of circuit breaker, the contacts were separated under oil without any form of arc control device. It was found that a significant increase in rated fault level could be obtained if the arc was enclosed by, what is now known as an 'arc control pot'. Initially, arc control pots were arranged to vent the gases produced by the arc axially in the direction of the moving contact. In effect, this was an 'explosion pot device'. Further increases in

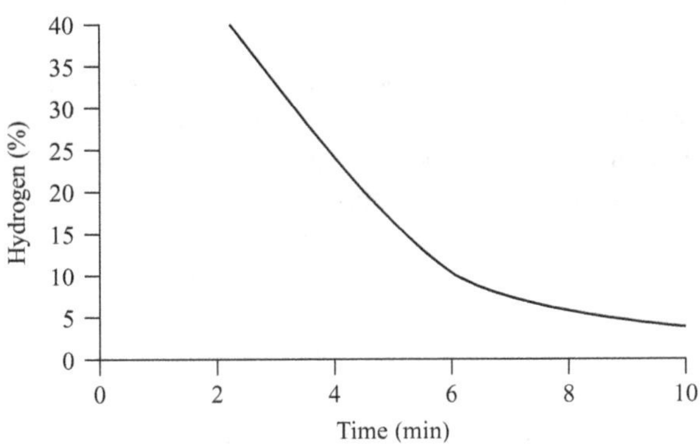

Figure 2.2 Dissipation time for hydrogen gas

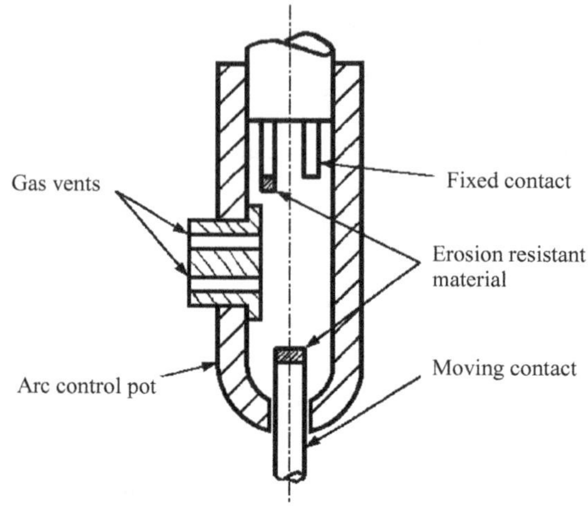

Figure 2.3 Diagrammatic cross-section of an arc control device

interrupting ability were realised when the gases were restricted from venting axially and instead were directed transversely across the arc channel. The relatively small internal volume of an arc control pot meant that very high internal gas pressures were generated during interruption of fault current. Figure 2.3 is a diagrammatic cross-section through an arc control pot. It will be seen that the pot encloses the fixed contact and has an orifice to allow the moving contact to enter. Both the fixed and moving contacts were fitted with erosion-resistant arc control tips to allow the contacts to meet the specified maximum number of six break operations at the fully asymmetrical fault current level without maintenance. It was the practice of some switchgear designers to arrange for a channel to be formed in the vent block to guide the arc into the most favourable position against the vents for interruption to take place.

Due to the very high gas pressures generated during interruption, a significant force was caused to act upon the end of the moving contact, which had to be controlled by dashpots within the contact drive system. This was to prevent an excessively long arc being created which would, in turn, create even greater forces on the moving contacts. As arc interruption was the result of the gas pressure generated within the arc control pot, it followed that the higher the current, the greater the pressure and the more efficient the interrupter became. However, the highest pressure that could be tolerated was dictated by the strength of the arc control pot and, for a given volume, was a function of the vent area. It also followed that arc control at the lowest fault level was much less efficient for a given arc length, as a lower gas pressure would be generated. The inherent danger was that the arc could be drawn outside the pot and, therefore, become out of control. A balance had to be achieved between the performance at the highest and lowest fault levels by optimising the vent area. This was an iterative experimental process, which frequently led to spectacular failures and left any young witnessing engineers with a healthy respect for switchgear which would remain with them for life.

The high gas pressure inside of the arc control pot also leads to another effect, which is plunger bar imbalance. It was found that if a single-phase high-level fault occurred in the phase furthest from the operating mechanism, the imbalance of forces could lead to the circuit breaker experiencing mechanical problems. This was recognised by the testing authorities, and consequently, they introduced a single-phase test to prove that the circuit breaker could cope with the imbalance. This test remains in today's mandatory test schedule and is applied to circuit breakers regardless of the type of interrupting technique.

Oil circuit breakers were often arranged to have two sets of contacts per phase connected in series to increase the interrupting ability and to help with mechanical balance. A cross-section through a typical double-break oil circuit breaker is shown in Figure 2.4.

The increase in breaking capacity achieved by employing two breaks in series per phase will not be twice that of a single break device, owing to the relative capacitance to earth giving unequal voltage sharing.

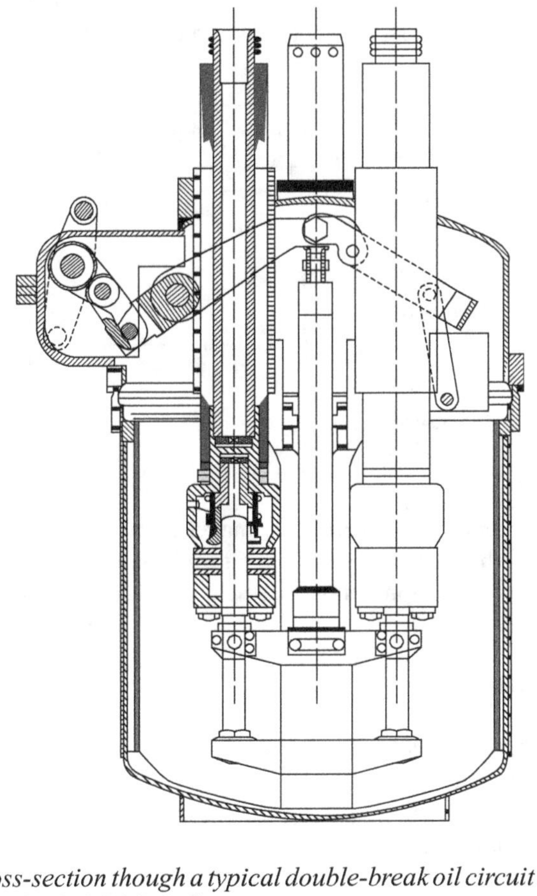

Figure 2.4 Cross-section though a typical double-break oil circuit breaker (courtesy of ALSTOM T&D Ltd)

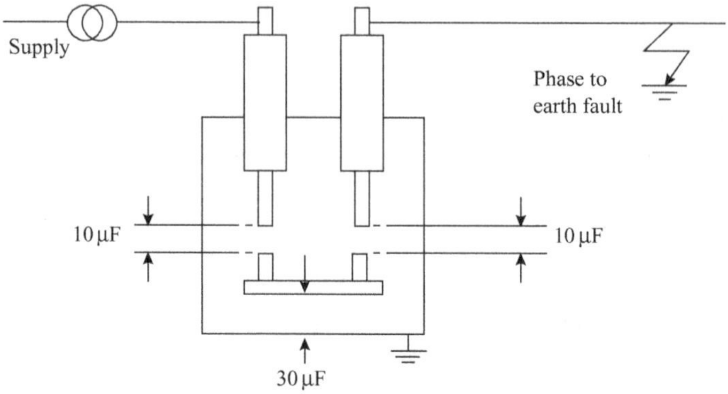

Figure 2.5 Diagram showing a phase to earth fault and the capacitance values across the contact gaps and to earth

Consider the diagram shown in Figure 2.5. As both the circuit breaker tank and the fault are at earth potential, a capacitance diagram can be constructed and the values resolved as shown below:

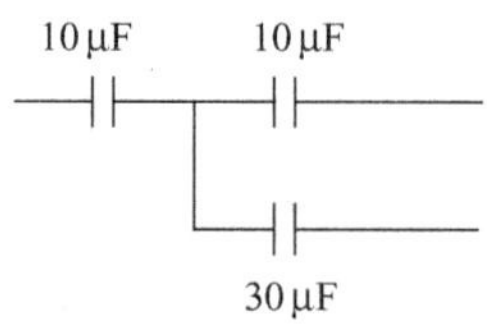

This resolves to:

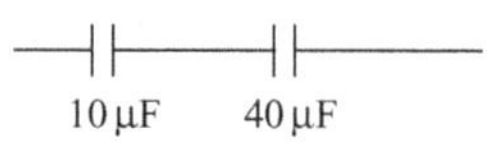

Therefore the voltage sharing across each gap will be:
80% and 20%

The second contact gap, therefore, only makes a small contribution to the circuit breaker performance but it also produces the same amount of gas and produces the same amount of contact erosion, but it does help to balance the mechanical forces.

2.4 Vacuum switchgear

The relationship between voltage withstand and electrode spacing is given by Paschen's Law. This suggests that the voltage withstand of a gap between electrodes is proportional to both the electrode spacing and the gas pressure. It is fortunate that this law is only true within finite limits, otherwise vacuum switchgear could not exist. Starting at atmospheric pressure, as air pressure is reduced, the voltage withstand also reduces, obeying Paschen's Law. However, at very low pressures, a remarkable

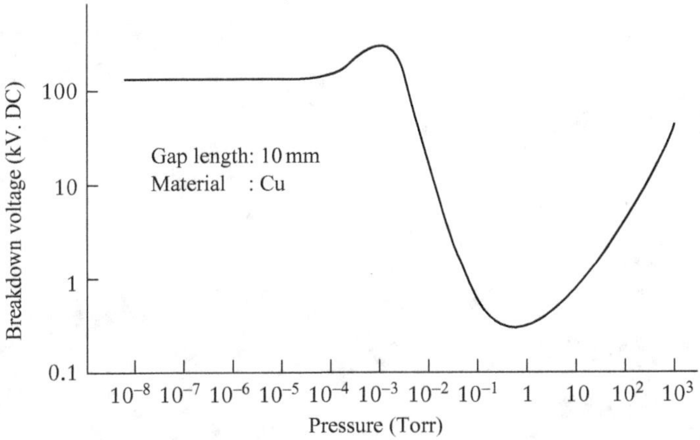

Figure 2.6 Paschen's curve at low pressures [4]

change takes place. Further reductions in pressure result in the withstand voltage increasing (see Figure 2.6).

Since the first commercial introduction of vacuum interrupters in the 1970s, continuous development has dramatically reduced the size and increased the short-circuit ratings available. The photograph in Figure 9.5 (Chapter 9) graphically demonstrates the changes that have taken place. The principle involved in vacuum interruption is very old, dating from Rittenhause's patent of 1893. However, the realisation of the practical working interrupter must rank alongside many of the great achievements in engineering. An arc cannot exist in a vacuum and requires metal vapour from the metal contacts to sustain itself, ideally until a natural current zero is reached. At this point, the metal vapour should condense back onto the contacts, denying conductivity so that current ceases to flow. Therefore, the contact materials are all important to the interrupting process. In addition, the materials used for the contacts must have the right characteristics for the conduction of normal current and they must minimise the natural tendency of metals to cold weld when pressed together under high-vacuum conditions. Further, they must not release gas when interrupting current, as this would destroy the high-vacuum necessary for the whole process to be repeated many times over, during the life of the vacuum interrupter.

It follows that, as a voltage will be impressed across an interrupter following current interruption, insulating materials have to be included in the design of the vacuum interrupter envelope. These insulating materials must be protected from condensing metal vapour from the contacts which would otherwise destroy their insulating properties. In practice, this is achieved in several different ways. Figure 2.7 is a photograph showing a sectioned vacuum interrupter. Protection for the internal surfaces of the insulating envelopes is provided by three metal shields, known as spatter shields, brazed to the centre band and end caps of the interrupter.

An alternative method of protecting the insulating envelope is to have both the fixed and moving contacts arranged to have their contact faces located within a central

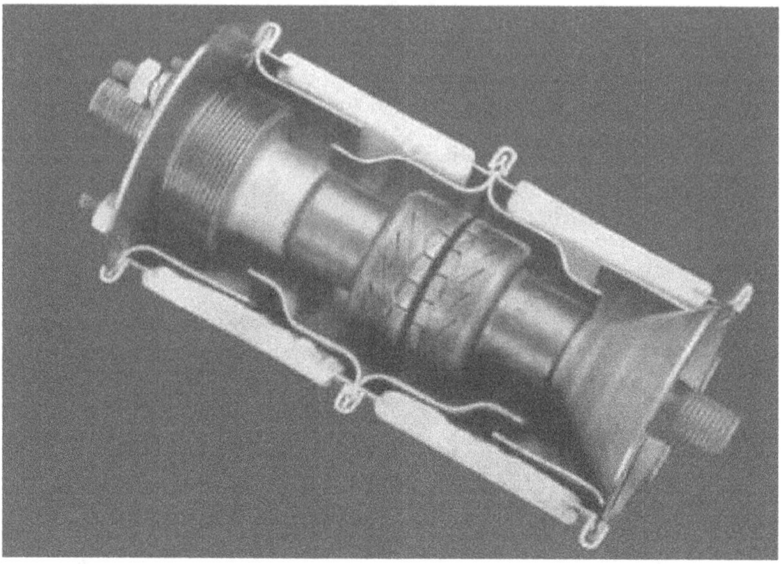

Figure 2.7 A sectioned type V801 vacuum interrupter (courtesy of ALSTOM T&D Ltd)

canister. The inside face of this canister acts in the same way as a spatter shield, in that condensing metal contact vapour is collected on its inner face, well away from the interrupter barrel insulating material. In this design of interrupter, the envelope is in the form of a barrel brazed to each end of the central canister. If a vacuum interrupter is cut open after a large number of fault current interruptions, the spatter shield will be found to have a copper plated appearance on its inner face, and the insulating materials forming the body of the interrupter should be clean.

As the vacuum interrupter contacts have to open and close within a vacuum envelope, it follows that the mechanical drive to the moving contact has to be able to conduct movement into the vacuum envelope through a gas-tight seal. In practice, this is done by arranging for the moving contact to be attached to the end plate of the vacuum interrupter by metal bellows. These bellows are usually manufactured from stainless steel which is either hydroformed to form the convolutions, or they are manufactured by welding the edges of a number of belled annular stainless steel discs, such as is shown in Figure 2.7. Regardless of the method of manufacture, the integrity of the bellows is of paramount importance. They must be able to maintain an internal vacuum over many years and many operating cycles. Therefore they must be tested over a very large number of operating cycles to ensure that they will not fail due to metal fatigue.

When interrupting currents of less than about 10 kA peak, the arc that is drawn between the contacts of a vacuum interrupter will be in the form of a number of parallel arcs. This is known as a diffuse arc and high-speed photographs show this to be like an internally illuminated cloud with a large number of points of light dancing across the contact surface. In reality, this form of arc consists of a large number of small parallel arcs that are kept separated from each other by electromagnetic force. This is because each arc acts like a small magnet and the arc roots simulate the magnetic poles. The poles of these arcs will, therefore, exert a repelling force on each other maintaining the arc in a diffuse state. At currents of about 10 kA and above, the main body of each of the small arcs will exert sufficient attractive force to overcome the pole effect and tend to cause the small arcs to fuse together into one large arc. A large single arc will produce an extremely high temperature at the arc root, causing an excessive amount of contact material to be vapourised, and so limiting the short-circuit current interrupting capability of the vacuum interrupter.

To minimise this effect and hence increase the short-circuit current rating, some manufacturers force the arc root to move over the contact face, preventing excessive temperatures and material vaporisation at one spot. They do this by employing what is known as contrate contacts. A contact of this type appears in Figure 2.7 and is shown in Figure 2.8.

The surface of a contrate contact is provided with a number of slots which, by electromagnetic force, will impose a self-generated rotational drive to the arc, increasing the short-circuit rating of the interrupter.

In more recent times, it was realised that if each of the small arcs in a diffuse arc could have their magnetic polarity increased, they would continue to maintain the diffuse state by resisting the parallel current effect, and thus increase the short-circuit rating of the interrupter.

There are several different ways in which this has been achieved by manufacturers. One of these, which is patented by Cooper Power Systems, is used in the type VSAM interrupter for their Kyle vacuum autorecloser. The fixed and moving contacts in this interrupter are in the shape of a spiral, which causes the electromagnetic field of the short-circuit current, as it approaches the contact face, to produce a vector of magnetic field to reinforce the magnetic polarity of the small individual parallel arcs (see Figure 2.9). This technique for arc control is known as an axial magnetic field.

Toshiba introduced another very successful method of producing an axial magnetic field to maintain the arc in a diffuse state up to very high fault current levels. The construction of the contacts using this method is shown in Figure 2.10, which

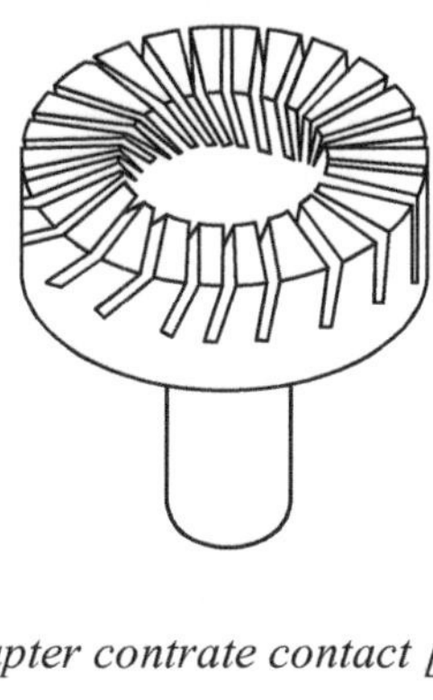

Figure 2.8 A vacuum interrupter contrate contact [4]

Figure 2.9 The type VSAM axial magnetic field interrupter (courtesy of Cooper Power Systems)

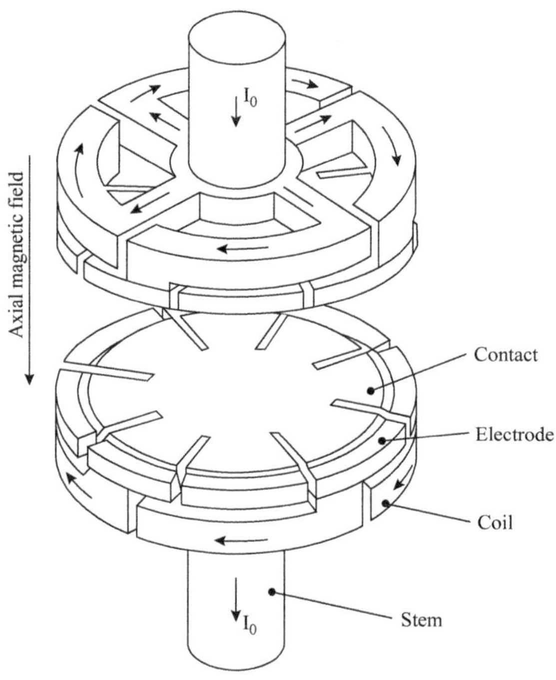

Figure 2.10 Contact arrangement providing an efficient axial magnetic field [5]

shows the current entering the contact arrangement via the top conductor stem. This current then flows outwards along the four radial arms, as indicated by the arrows. The current path changes when it reaches the periphery of the contact, which it follows for about 90°, where it connects to the contact interface. The current paths in the outgoing contact mirror those of the incoming contact and it is these paths which provide the strong axial magnetic field that maintains the arc in a diffuse state up to very high levels of fault current.

In order to be competitive, the manufacture of vacuum interrupters must be carried out in significant quantities. The manufacturing equipment is very specialised and, therefore, expensive. For example, consider how specialised the vacuum furnace which is used in the manufacture must be. The brazed joints, both metal to metal, and metal to insulating material, have to be carried out in such a furnace at the same time in the presence of a very high-vacuum. As heat convection cannot be used, the heat necessary for brazing can only be radiated and conducted to the joints, without causing overheating of some of the joints and subsequent loss of brazing material. Such an arrangement requires very careful design and is expensive to implement. In addition, the manufacturing conditions have to be such that no measurable contamination can be allowed on the internal components after full cleaning. This means that a clean room with positive internal pressure has to be provided. This clean room will require air locks for access of material and personnel. Such a room is shown in Figure 2.11.

After assembly, all vacuum interrupters are subjected to routine tests to ensure compliance with declared acceptance criteria. These tests will usually include a measurement of the internal vacuum which will be noted. The degree of vacuum will normally be of the order of 10^{-5}–10^{-7} torr. The vacuum interrupters will often be subjected to a second vacuum pressure measurement after a fixed elapsed time, which by comparing the two measurements, will allow calculations of leak rate to be made, and so confirm the manufacturers' published shelf life of the vacuum interrupter.

Figure 2.11 *Assembly of vacuum interrupters in a clean room (courtesy of ALSTOM Medium Voltage Switchgear, South Africa)*

2.5 SF_6 switchgear

At about the time that Rittenhause was filing the first patent on vacuum interrupters, work was being carried out in Paris that would lead to the creation of sulphur hexafluoride, SF_6. Two French scientists, Moissan and Lebeau produced the first samples of this gas at the turn of the century during laboratory experiments involving the electrolytic action of fluorine on sulphur in a copper tube. The gas produced, SF_6, was a remarkably stable gas that would rapidly recombine if dissociated. It consisted of a large central sulphur molecule surrounded by six fluorine molecules. This arrangement is shown diagrammatically in Figure 2.12.

The model shown in Figure 2.12 is a little misleading as the actual radius path of the electrons is much smaller than the diagram would suggest. Although the excellent electrical insulating properties of SF_6 were explored very soon after its discovery, the gas remained a scientific curiosity for many years. Apart from finding an application in X-ray apparatus in the 1930s, little was done to investigate its commercial prospects until it began to be produced in large quantities in the mid-1940s as a by-product of the nuclear industrial programme.

The excellent dielectric properties of SF_6 gas had suggested that it could be applied effectively to the insulation of extra high-voltage equipment. However, it is well known that there is little relationship between the dielectric strength of a gas and its ability to extinguish an electric arc. For example, hydrogen has only about half the dielectric strength of air but it has the ability to interrupt several times the current that air will interrupt under the same test conditions. It was, therefore, left to the Westinghouse Company of the USA to discover the remarkable interrupting ability of SF_6 gas and in the 1950s, they went on to produce the first commercial

Figure 2.12 The SF_6 molecule (courtesy of Solvay Fluor und Derivate GmbH)

SF_6 switchgear. At transmission voltage levels, the economic challenge to air blast switchgear was overwhelming and many other manufacturers started to research, design and produce SF_6 for their own transmission switchgear.

An SF_6 enclosure will need to be as gas-tight as possible, as the gas pressure used will influence the voltages that can be handled and the short-circuit current that can be interrupted. This will require rigorous gas tightness checks to be made. As SF_6 gas is a halogen, very sensitive gas leakage detectors can be used, which will detect very small leaks of gas. However, experience has shown that no measurable leak can be tolerated, as gas leaks will only get worse with time. In addition, leaks from welded joints in an enclosure can take up to 8 days to materialise, as the actual fault in the weld can be very small and the leak path to the measuring point can be large. Therefore, where welded joints are to form the gas envelope, it would be prudent to allow 10–14 days before declaring the enclosure as being gas-tight. Leaks past gas seals, usually caused by scratch marks on the sealing surface, can take up to 2 days to materialise. A typical halogen leak detector, which uses negative ion capture, is shown in Figure 2.13.

SF_6 gas, if released into the atmosphere, will contribute to global warming as it is a greenhouse gas. However, the contribution of SF_6 towards global warming is extremely small and the majority of SF_6 gas released into the atmosphere does not emanate from electrical switchgear. A recent analysis of greenhouse gases in the atmosphere (which contribute to global warming) is shown in Figure 2.14. Release of SF_6 gas into the atmosphere can occur when the gas is used in the casting process of magnesium and aluminium or as a filling medium for double glazed windows. In any case, the gas is far too expensive to be released from switchgear unnecessarily. Gas reclamation plant is commercially available which will not only protect

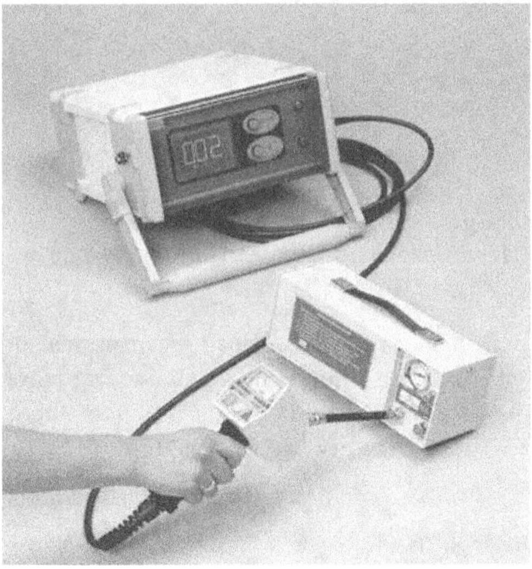

Figure 2.13 A halogen gas leak detector (courtesy of Ion Science Ltd)

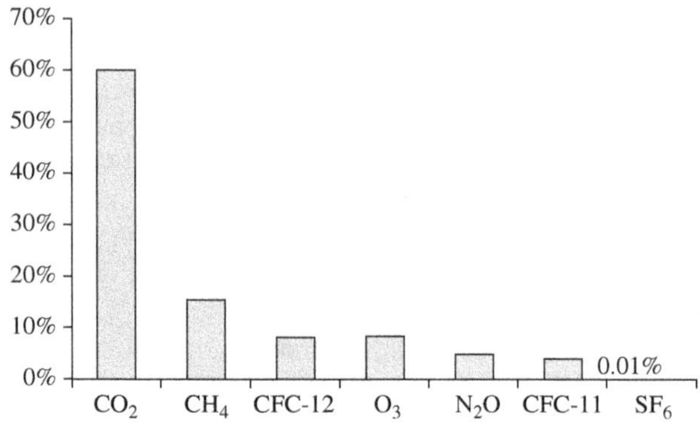

Figure 2.14 Histogram showing global warming contribution from various gases

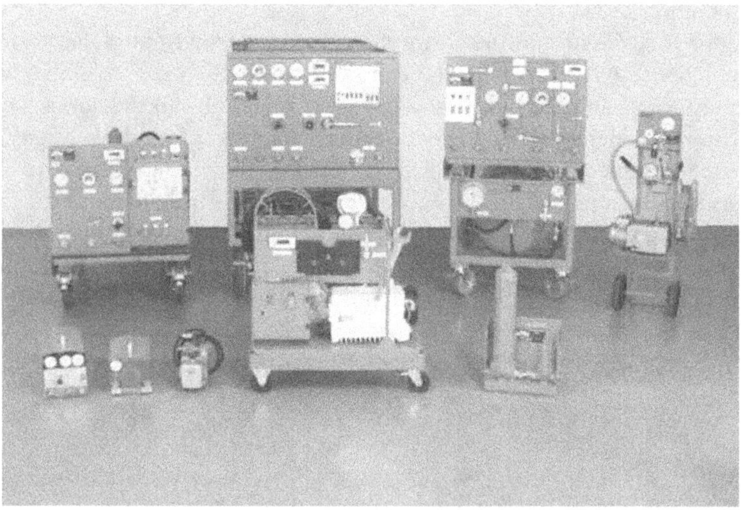

Figure 2.15 A selection of the range of SF_6 gas handling equipment (courtesy of DILO Armaturen und Anlagen GmbH Germany)

the atmosphere but will also pay for itself in time in terms of the cost of the gas saved. Examples of gas handling equipment are shown in Figure 2.15.

In order to design successful SF_6 switchgear products, there are certain areas in the design detail that will require very careful attention.

2.5.1 Gas pressure

The internal gas pressure must be such that the SF_6 gas will not start to change from its gaseous form into a liquid when at the minimum rated ambient temperature. This is

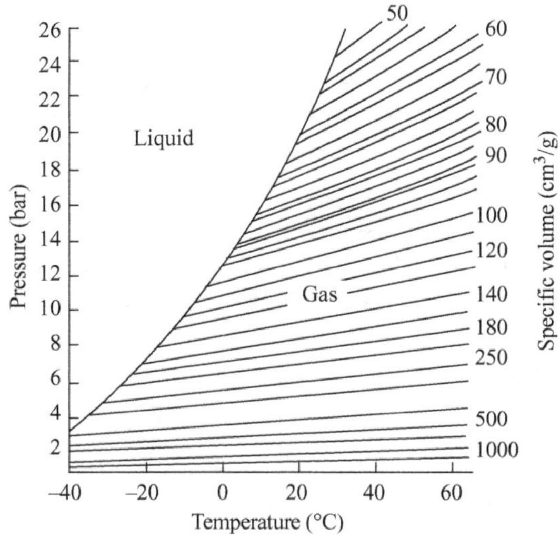

Figure 2.16 Pressure/temperature characteristics of SF_6 gas [6]

to prevent the liquefied gas running down insulation and causing any contaminating particles to line up, which was a problem in the early designs.

Consider the pressure/temperature characteristic of the gas, as shown in Figure 2.16.

Outdoor SF_6 switchgear has a minimum rated temperature of −25°C. This means that according to Figure 2.15, the maximum internal gas pressure that can be used is 5.9 bar (absolute).

2.5.2 Contact design

An electrical arc in SF_6 gas will cause the gas to dissociate. However, most of the dissociated products will recombine into SF_6 as the gas cools. A small percentage will combine with vapour from the arcing contacts to form metallic fluorides in the form of a finely divided grey powder. These metallic fluorides are insulators, making it necessary for separable contacts in arced SF_6 to be of the wiping, self-cleaning type. Butt-type contacts cannot be used as these would not have surfaces free of metallic fluorides and would, therefore, be the cause of high resistance, and produce a higher temperature rise than would otherwise be the case.

2.5.3 Gas dryness

The metallic fluorides discussed earlier are, in reality, the salts of acids and must be kept completely dry to prevent the formation of acids, which would lead to subsequent insulation failure and corrosion. The assumption that if a unit is gas tight and internally dry it will remain so is false. Under partial pressures, moisture will look on the gas envelope as containing a vacuum and will try to ingress, usually through gas seals that include rubber materials having a degree of permeability. In practice, the dryness of the gas is controlled by the choice of materials and by the inclusion of molecular sieves, such as sodium alumino-silicate. This latter material exerts a strong attraction to moisture and is able to dry out a gas which is already very dry.

2.5.4 Choice of materials

The inclusion within SF_6 gas of materials having a significant moisture content, such as certain grades of Nylon, will have two effects. The first is that these materials will give up their moisture to the gas, which as already described can be dangerous, and the second is that the material will lose a substantial measure of its mechanical strength. All materials, therefore, should be examined for moisture content and selected with this in mind.

2.6 Interrupter types

2.6.1 Rotating arc SF_6 interrupters

The earliest practical SF_6 transmission circuit breakers stored the gas stored at a higher pressure and released this through a blast valve to a lower pressure chamber to extinguish the fault current arc. The complication of this construction was the reliance upon gas heaters to prevent the gas liquefying, and the inclusion of an internal gas pump to return the gas to a high-pressure chamber. These complications and costs led to the introduction of the much simpler construction of 'Puffer interrupter' which in later years was to be used within distribution switchgear.

The extension of the use of SF_6 gas from transmission switchgear voltages to distribution switchgear voltages was not simultaneous, and lagged by almost ten years. The reason was probably due to switchgear designers being dedicated to one or the other of the types. However, in the early 1980s, South Wales Switchgear introduced the type Hawkgas 12, which was interchangeable with its previous range of indoor vertically isolated oil switchgear, and at about the same time Brush switchgear introduced the Falcon ring-main unit and in 1982 the type PMR autorecloser. All three of these designs used a rotating arc interrupter. The physical arrangement of the fixed and moving contacts together with the interrupter coil is shown in Figure 2.17. It will be seen that the interrupter consists of simple separable contacts adjacent to an interrupter coil.

In Figure 2.17, the moving contact is hinged and rotates about its axis pin (4) from the fully engaged position within the fixed contact (3) to the fully open position, concentric with the axis of the interrupter coil (2). The fixed contact has an extended arcing finger which is arranged to be the last point of contact with the moving contact during a contact opening operation. The interrupter coil (2) consists of a copper coil having one end terminated on the coil former and the other connected electrically to the fixed contact. The principle of operation is shown in Figure 2.18.

It can be seen that an arc will be drawn between the fixed and moving contacts as the moving contact is driven towards its open, central position, co-axial with the

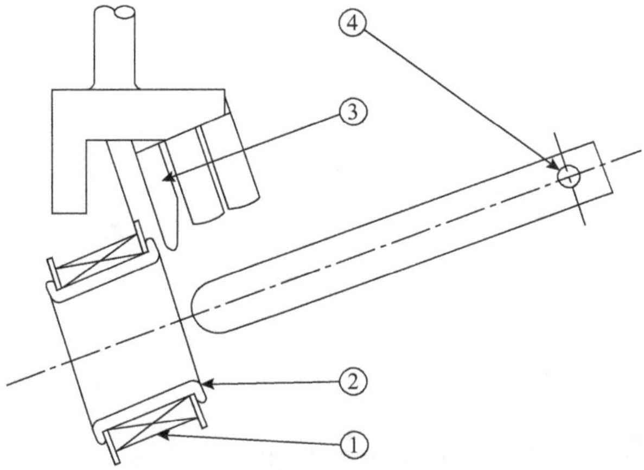

Figure 2.17 An arrangement of the contacts and rotating arc SF_6 interrupter [7]

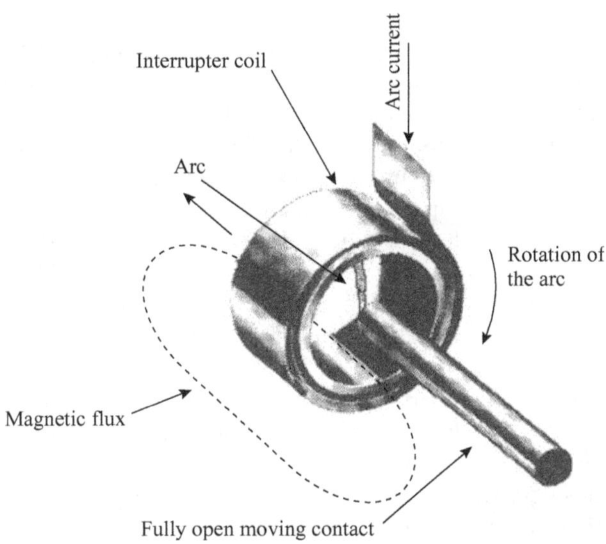

Figure 2.18 Principle of operation of a rotating arc SF_6 interrupter [7]

interrupter coil. This coil is usually in the form of a thin copper strip and, as the voltage drop across each turn of the coil is small, as when conducting fault current, the interturn insulation is usually in the form of a thin Melinex tape.

The arc root at the fixed contact is electromagnetically driven to transfer onto the coil former and the fault current is then forced to flow through the coil, producing a magnetic field at right angles to the arc. This field causes the arc to be driven rotationally around the inside of the coil former, by the same principle as that of

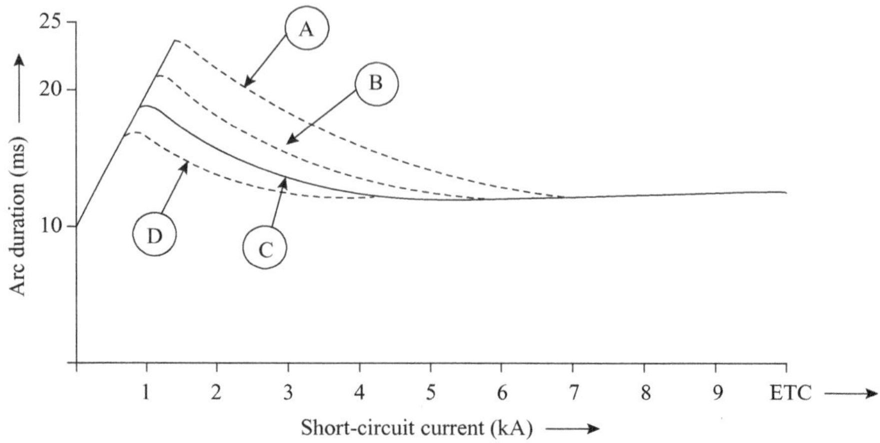

Figure 2.19 Effect of interrupter coil turns on arc duration

an electrical motor, bringing it at speed into cool gas, leading to rapid fault current interruption. Usually this interruption takes place at the first available current zero when interrupting the rated short-circuit current. However, the rotating arc interrupter has a limited capability in terms of the magnitude of the short-circuit current that it is capable of handling. At the peak currents associated with a fault current of about 27 kA rms, the coil former will collapse into a wine glass shape due to the electromagnetic crushing force imposed by the coil windings.

The arc duration at low fault current levels will be influenced by the number of turns of foil on the interrupter coil. The greater the number of turns, the shorter the arc duration. This is shown in Figure 2.19. This figure shows the effect of the number of coil turns on the arc duration at the rated short-circuit current. It will be seen that each characteristic is in two parts. The initial part is a straight line increase in arc duration with current and this is followed by a reduction in arc duration to an almost constant value, independent of the fault current. This is because the electromagnetic drive on the arc is very weak at low currents and then starts to become effective as the current levels being interrupted are increased.

The switchgear designer has to compromise between the number of coil turns, the peak current to be handled and the arc duration. Too many turns will give early control of the arc duration but will limit the maximum fault current that can be handled. The four curves shown in Figure 2.19, A, B, C, and D, indicate the effect of increasing the number of coil turns and suggest that the designer compromised by selecting curve C. The actual variation in number of turns is relatively small and is usually between 17 and 25.

In practice, the moving contact can either start from a fixed contact at the edge of the coil, as in the designs used by South Wales Switchgear Ltd and Brush Switchgear Ltd, or start from a fixed contact located on the axis of the interrupter coil, as favoured by Groupe Schneider.

2.6.2 The SF_6 puffer interrupter

The SF_6 puffer interrupter was initially developed for high-voltage switchgear, typically 145 kV and above, and is almost universally used at these voltages today. The name 'puffer' is deceptive, as it does not convey the power of this technique of fault current interrupter. In fact, the gas issuing through a well designed interrupter nozzle will be travelling at the speed of sound. Typical features of an SF_6 puffer interrupter are shown in Figure 2.20. As can be seen in this diagram, a typical SF_6 puffer interrupter consists of a moving contact which has a cylinder attached that is designed to operate against a fixed piston. At distribution voltage levels, the physical construction of the SF_6 puffer interrupter can vary depending upon the design philosophy of the manufacturer. For example, it is quite common for the puffer cylinder and/or the piston to be formed as part of an integral insulating moulding or mouldings. During the opening stroke, the gas within the cylinder is compressed and has a very restricted flow through the throat of the insulating nozzle until the nozzle clears the fixed contact.

The downstream divergence angle of the nozzle is important in order to obtain supersonic gas flow and maximise interrupting capability. The construction is essentially very simple in that there is only one moving part; however, there are minor

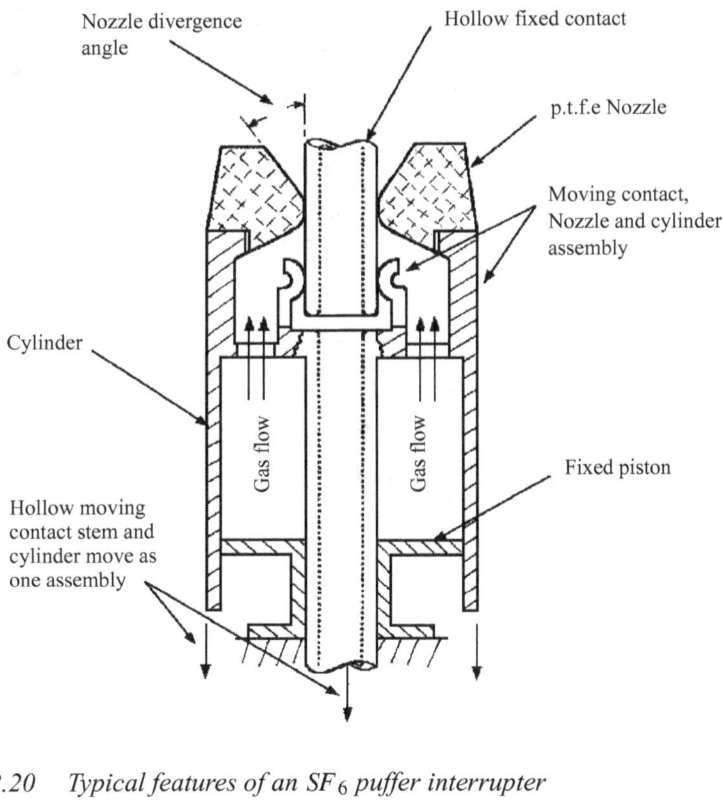

Figure 2.20 Typical features of an SF_6 puffer interrupter

Table 2.1 Comparison of the features of vacuum and SF_6 switchgear

Feature	Vacuum switchgear		SF_6 switchgear	
	Advantages	Disadvantages	Advantages	Disadvantages
Long contact life when closing on to, and breaking fault currents	Yes	–	Yes but not usually as long as vacuum	–
Freedom from fire hazard	Virtual freedom from fire hazard	–	Virtual freedom from fire hazard	–
Consistent arcing and total clearing time	Yes	–	Puffer is consistent	Rotating arc may show an increase at lower currents
Restrike free operation	Most types are restrike free	–	Most types are restrike free	–
High-speed fault clearance	Most types have short total clearance times	–	Puffer types have short total clearance times	Rotating arc may show an increase at lower currents
Little or no overvoltage generation	–	Susceptable to current chopping and re-ignition surges but within normal distribution applications the values reached do not require any special measures to be taken	Rotating arc provides soft interruption without overvoltage generation	Certain puffer types will chop low levels of current
No exhaust gases	No exhaust gases	–	No exhaust gases	–
Minimal maintenance	Low maintenance	–	Low maintenance	–
Simple operation	Yes	–	Yes	–
Minimal space	Normal with air clearances	–	Advantage can be taken of the gas insulation to dramatically reduce dimensions	–
Freedom from environmental effects	–	Air filled chambers could present difficulties in high humidity and temperature swings	The gas enclosure ensures freedom from environmental effects	–
End of life disposal	No particular problem	–	–	Will require the services of specialist companies

complications. For example, the fixed piston has to be fitted with a non-return valve to allow the cylinder to fill with gas during a closing stroke, so that a full charge of gas is available for a subsequent break operation, in case the circuit breaker closes onto a fault.

The SF_6 puffer interrupter has the advantage of not suffering from the limitation in fault current that was described as affecting the rotating arc interrupter. However, there are two features that must be recognised. The first is that during interruption, the gas pressure within the interrupter cylinder will be significantly increased, which will tend to stall movement as the operating mechanism has to overcome this gas pressure. In this regard, it is exactly opposite that of the oil circuit breaker, which tends to accelerate the moving contacts towards the open position. The SF_6 puffer interrupter, therefore, requires a relatively large input of mechanism energy when compared to the rotating arc interrupter, which suffers no feedback of energy from the arc. The second feature that should be recognised is that the blast of gas will only be available for a short finite time and that interruption can only take place in that period, no matter when current zero occurs. The SF_6 puffer interrupter, therefore, has to be proven to have an interrupting window that will embrace at least one current zero. By way of an example, if during short-circuit tests it is found that the minimum arc duration before interruption is 3 ms, the testing authority must be satisfied that the interrupter will also clear the fault current at 13 ms on a 50 Hz system.

2.6.3 The relative merits of vacuum and SF_6 interrupters (Table 2.1)

Both vacuum and SF_6 switchgears are produced in many forms, as well as some which are hybrids, for all applications within the gambit of distribution switchgear. These include primary and secondary substations, indoor, outdoor, pad mount and pole mounted forms. The designs include dead tank, metalclad and live tank. The choice is wide and is made by the user on operational, economical and technical grounds.

Chapter 3

Fault level calculations

The result of a fault in the electrical distribution network can be relatively minor, as shown in Figure 3.1 where a dry-type cable termination was incorrectly fitted, or it can be catastrophic, as shown in Figure 3.2. The degree of damage depends on the impedance of the circuit carrying the fault current.

From this, it will be appreciated that when a short-circuit fault occurs in a network, such as that shown at point 'A' in Figure 3.3, the resulting short-circuit current will only be limited by the elements of impedance that are remaining in the circuit.

At medium and high voltages these remaining elements are highly inductive, and have a much smaller value of impedance than the load that was previously being supplied. The resulting current will, therefore, be considerably higher than the full

Figure 3.1 Electrical fault in a dry-type cable termination

Figure 3.2 Catastrophic failure within a substation

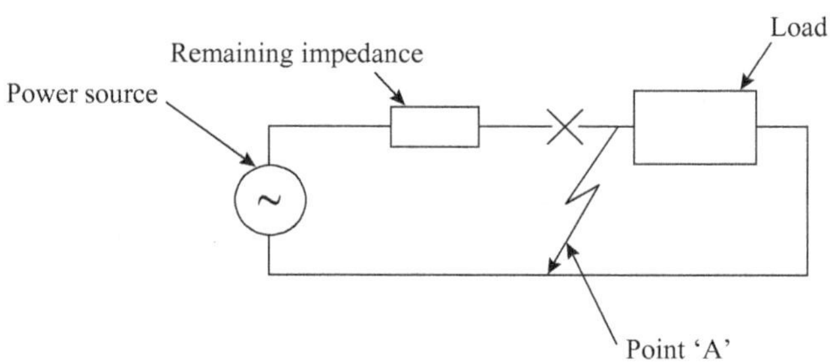

Figure 3.3 Basic short circuit

load current previously being supplied, and the power factor, considerably lower. For example, a circuit that was supplying 630 A normal current load at 0.8 power factor, could suddenly experience a current increase to 31,500 A with a power factor of 0.07 and if the short-circuit rating of the connected plant was not equal to, or higher than 31,500 A, catastrophic failure could be expected.

When planning a new installation, or modifying an existing installation by adding in a transformer or making a cable connection to another substation, the effect on the fault level needs to be determined. This is to ensure that the installed plant will still be within its rating and will be able to carry and interrupt the fault current safely. There are a number of specialist companies, and some software packages available, to carry out these calculations but a practising engineer should be able to determine the likely fault level at the feasibility stage.

The calculation of short-circuit currents is made easier by expressing the remaining elements of impedance in terms of their 'per unit' (pu) values. A definition of which is: 'the pu value is the voltage drop due to the element of impedance when passing full short-circuit current and is expressed as a fraction of the full load voltage'. This definition is worth committing to memory. The influence of the resistance of the remaining circuit elements at medium and high voltage levels is minimal and can be ignored. This also makes fault level calculations much simpler, avoiding vectorial solutions to the inductance and resistance elements. Resolving the pu reactances remaining in the circuit is carried out in the same way that interconnected resistances are resolved, which is as follows:

For series connections:

$$R_{\text{total}} = R_1 + R_2 + \text{etc.}$$

For parallel connections:

$$\frac{1}{R_{\text{total}}} = \frac{1}{R_1} + \frac{1}{R_2} + \text{etc.}$$

Before carrying out fault level calculations, it is recommended that a diagram of the circuit impedances be drawn, as this will help in visualising and resolving a network. This diagram should have high voltage as a top horizontal line, and earth as a bottom horizontal line, with the circuit network connecting the two. A diagram like this is normally useful but becomes essential when resolving networks having complex interconnections.

An example of a very simple fault level calculation follows, where there is only one element of impedance. You will notice that in the calculations the network impedance upstream from the transformer is ignored, as this will have a very small value.

Example 3.1

A 200 MVA, 11 kV, three-phase transformer has an Xpu of 1.5. What is the maximum fault current that will flow in the event of a short circuit occurring on its outgoing terminals (Figure 3.4)?

$$\text{MVA}_{\text{sc}} = \frac{\text{three-phase MVA}_{\text{plant}}}{\text{Xpu}} = \frac{500}{1.5} = 333.3\,\text{MVA} = \sqrt{3} \times V \times I_{\text{sc}}$$

Therefore

$$I_{\text{sc}} = \frac{333.3}{\sqrt{3} \times 11} = 17.5\,\text{kA}$$

More complex systems, having different voltages between the source and the fault, may appear daunting. This is because the voltage drop is proportional to current.

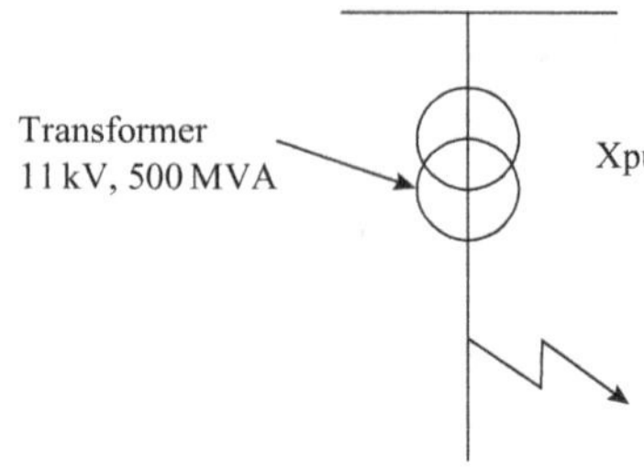

Figure 3.4 Fault current limited by transformer impedance

Therefore the pu reactance is only valid at the rated current. This is overcome by adopting a common base MVA and converting the actual plant MVA to that base. It is important to understand that the adoption of a common base does not affect the result. The value chosen for the MVA_{base} is usually one that minimises the conversion calculations. The procedure to follow when resolving complex networks is:

(1) Adopt a base MVA.
(2) Refer each reactance to the adopted base by using:

$$Xpu = \frac{X_{plant}}{MVA_{plant}} \times MVA_{base}$$

(3) Simplify the system component reactances to calculate the total pu reactance.
(4) Calculate the fault MVA using:

$$MVA_{sc} = \frac{MVA_{base}}{MVA_{plant}}$$

Finally,
(5) Calculate the short-circuit current from the fault MVA using $MVA = \sqrt{3} \times V \times I_{sc}$. Therefore,

$$I_{sc} = \frac{MVA_{sc}}{\sqrt{3} \times V}$$

An example of calculating the fault level in a system involving more than one voltage is as follows (Figure 3.5):

Example 3.2
Two 60 MVA, 0.2 pu generators feed an overhead 132 kV transmission line through a single 11 kV/132 kV, 120 MVA, 0.1 pu transformer. Calculate the fault current at the circuit breaker if a short circuit occurred on the overhead line connected to the outgoing terminals of the 132 kV transformer. Ignore the impedance of the cables connecting the generators to the transformer and assume that the impedance of the overhead line between the transformer and the fault is 1 Ω.

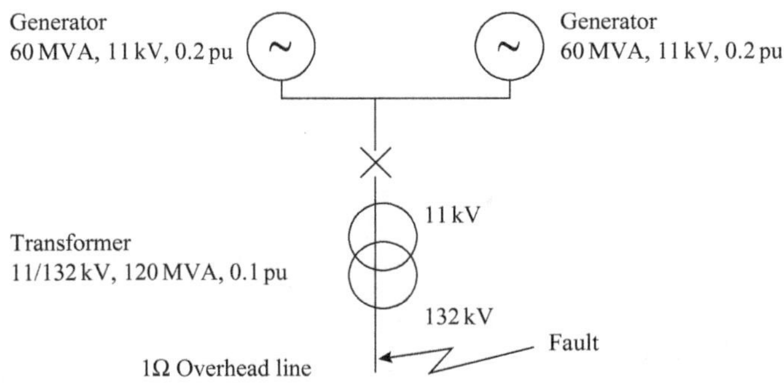

Figure 3.5 Example with generators, a transformer and more than one voltage

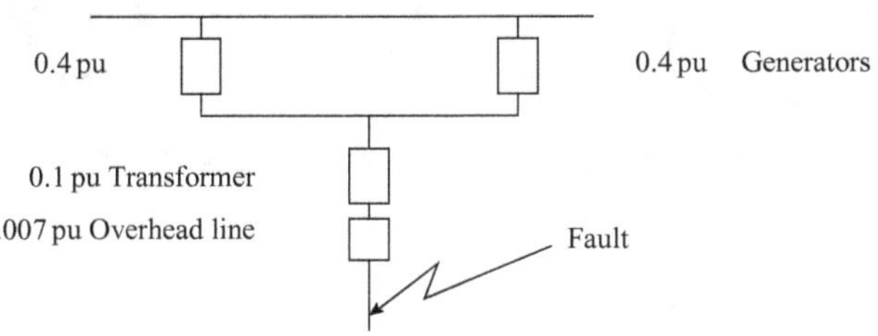

Figure 3.6 Diagram of the network

For convenience, adopt a base MVA equal to that of the transformer, that is, 120 MVA:

$$\text{Xpu of overhead line} = \frac{1\,\Omega \times \text{MVA}_{\text{base}}}{V \times V} = \frac{1 \times 120}{132 \times 132} = 0.007.$$

The pu reactance of each generator, converted to the adopted base MVA is then calculated from:

$$\text{Xpu base} = \frac{\text{X}_{\text{phase}} \times \text{MVA}_{\text{base}}}{\text{MVA}_{\text{plant}}} = \frac{0.2 \times 120}{60} = 0.4\,\text{pu}.$$

As the transformer has an MVA equal to the adopted base, its pu remains at 0.1 pu. The system can now be re-drawn in order to simplify and resolve the pu reactance values (Figure 3.6).

The reactance values can be simplified and resolved in the same way as resistance values and are as shown in Figure 3.7.

The total pu reactance value Xpu is

$$\text{Xpu} = 0.2 + 0.1 + 0.007 = 0.307.$$

The short-circuit MVA can then be calculated as:

$$\text{MVA}_{\text{sc}} = \frac{\text{MVA}_{\text{base}}}{\text{Xpu}} = \frac{120}{0.307} = 390.1\,\text{MVA}.$$

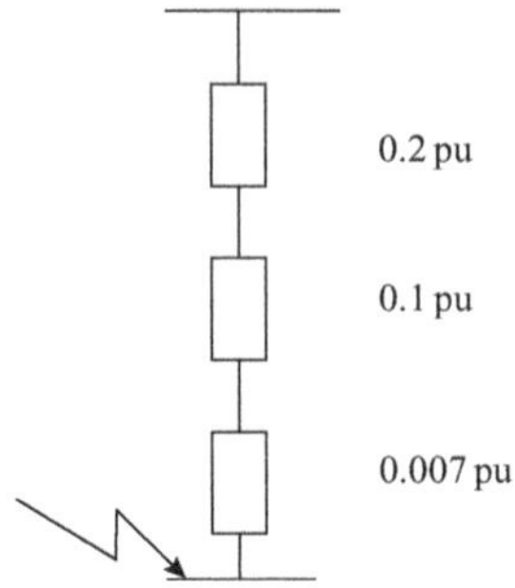

Figure 3.7 Reactance values simplified and resolved

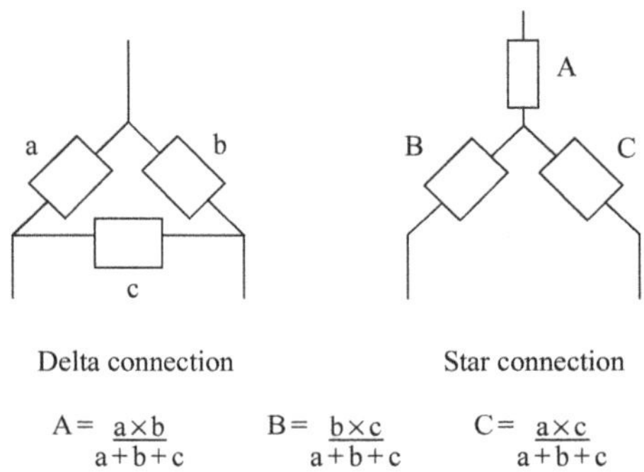

Figure 3.8 Delta–Star transformations

Finally the actual short-circuit current that will flow as a result of the fault can be calculated as:

$$I_{sc} = \frac{\text{MVA}_{sc}}{\sqrt{3} \times V} = \frac{390.1}{1.732 \times 11} = 20.48\,\text{kA}.$$

3.1 Impedance resolution within complicated networks

So far, we have seen how to determine the fault level within networks of equal and unequal voltages. There is one further condition that needs to be studied in order to be able to calculate fault levels in all types of networks. This final consideration is the resolution of faults within interconnected networks having impedances in the form of a Delta connection which, without translation into a Star connection, cannot be resolved. The Delta connection is not always easily recognised at the outset, but it does frequently occur, and becomes obvious when a diagram of circuit impedances is drawn and simplified. Figure 3.8 illustrates the Star and Delta types of impedance

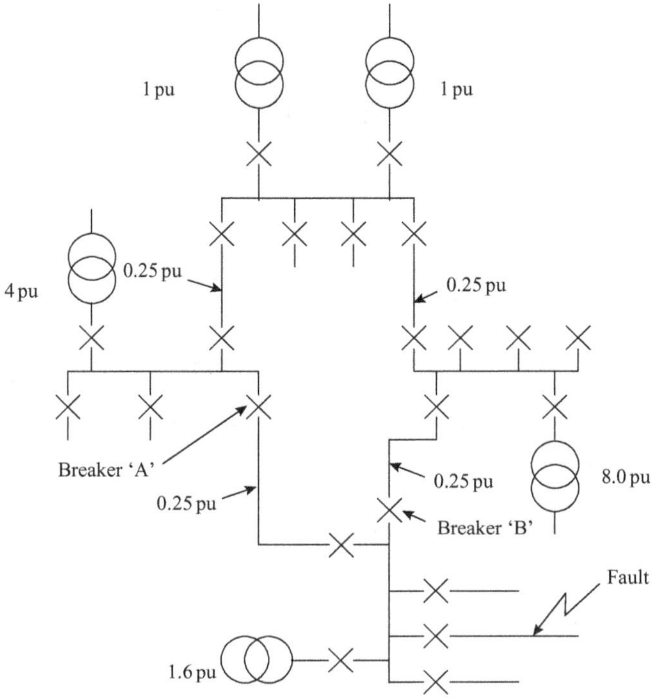

Figure 3.9 A complex network

connections and the relatively simple calculations that have to be made in order to allow the impedance diagrams to be simplified.

It can be seen that the Delta connection cannot be directly resolved into series and parallel impedances for circuit simplification and resolution. In order to overcome this problem, a Delta to Star transformation has to be carried out. The Star equivalent impedance of the Delta configuration can be calculated as shown. The following includes examples that require Delta–Star transformations to arrive at the answer.

Example 3.3

Five 11 kV primary substations are interconnected as shown in Figure 3.9. The pu values given are to a 100 MVA base. Calculate the fault level at the point indicated assuming that all circuit breakers are closed.

We start by representing the network graphically (Figure 3.9), as in Figure 3.10, showing the pu values at the adopted base value of 100 MVA.

This simplifies to Figure 3.11.

It can be seen that the 4.0, 0.5 and the 0.25 impedances are Delta connected and will have to be converted to a Star connection in order to further simplify the diagram. The Delta element is shown in Figure 3.12.

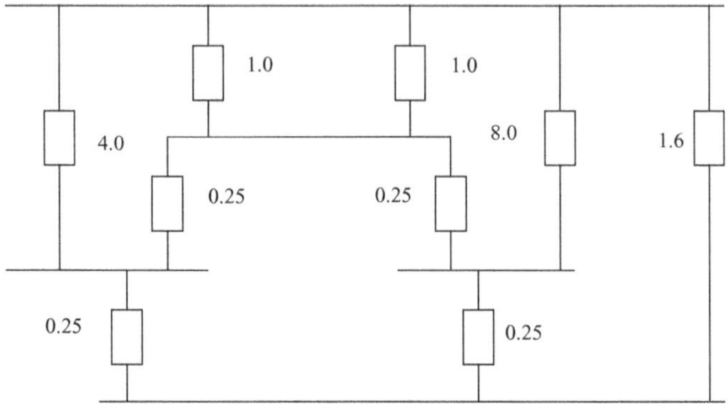

Figure 3.10 Diagrammatic representation of the network

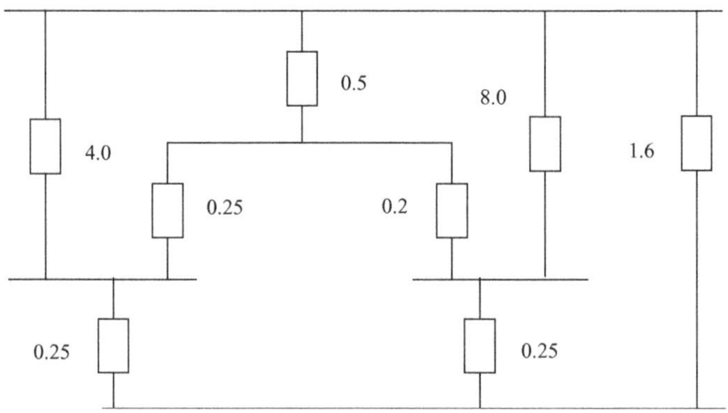

Figure 3.11 Simplification of the network diagram

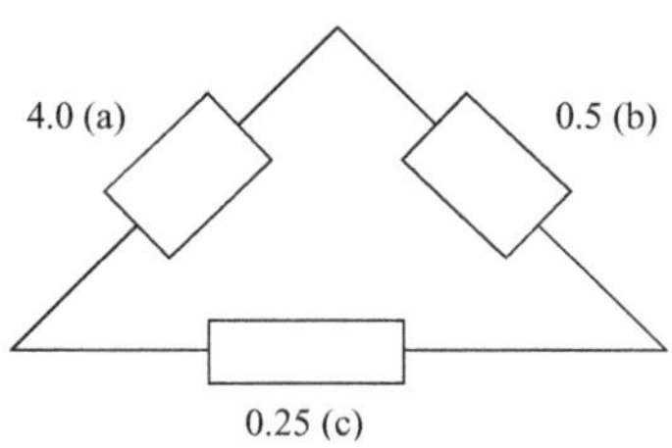

Figure 3.12 The Delta element within the network

As shown earlier, the equivalent limbs of the Star connection are calculated by dividing the product of adjacent sides by the sum of all three sides. The equivalent Star connection is then as given by Figure 3.13.

This further simplifies the network to that shown in Figure 3.14. This can now be further simplified, as shown in Figure 3.15.

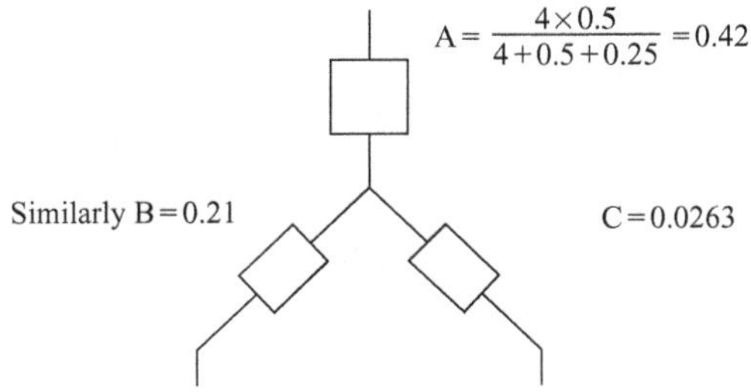

Figure 3.13 Star connected equivalent to the Delta element in the circuit

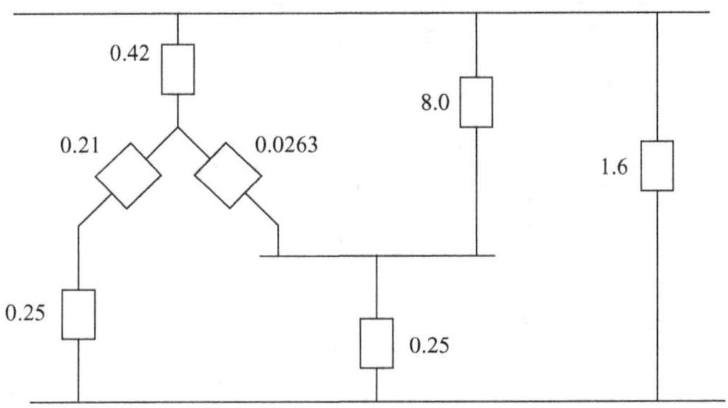

Figure 3.14 Further simplification of the network

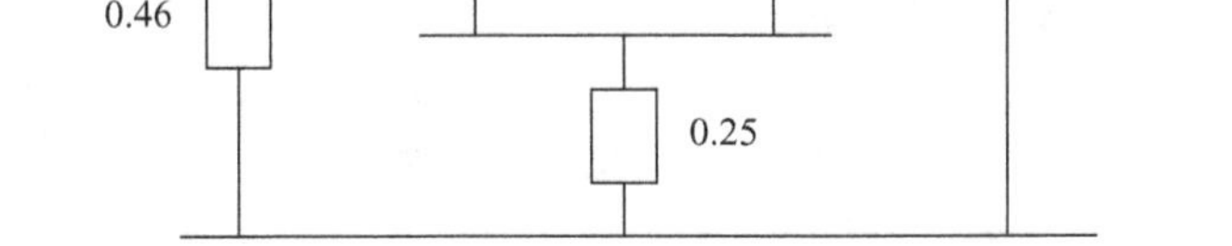

Figure 3.15 Further resolution of the network

It will be seen that impedances 0.42, 0.2263 and 8.0 form another Delta connection which has to be transformed into a Star connection to allow the impedance of the network to be resolved (Figure 3.16).

The network now changes to that shown in Figure 3.17. This can then be simplified to Figure 3.18. These networks can be further simplified as shown in Figures 3.19 and 3.20.

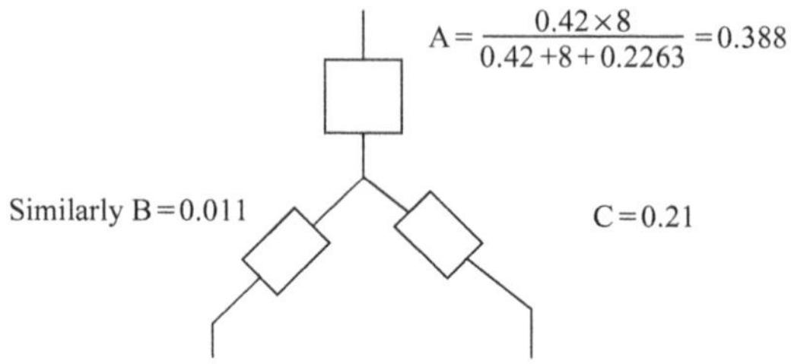

Figure 3.16 Transformation of the Delta element into a Star connection

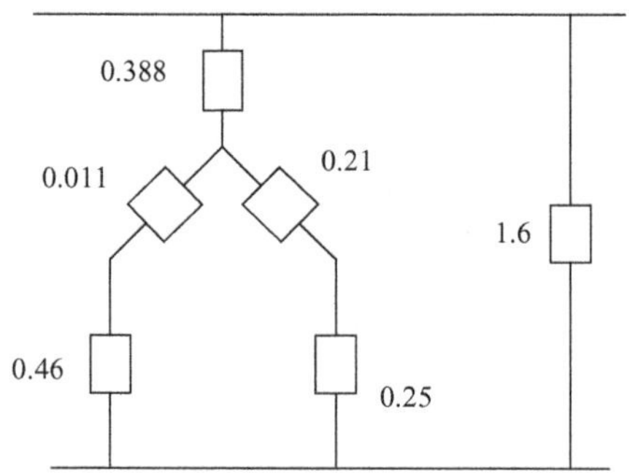

Figure 3.17 Network with Star element

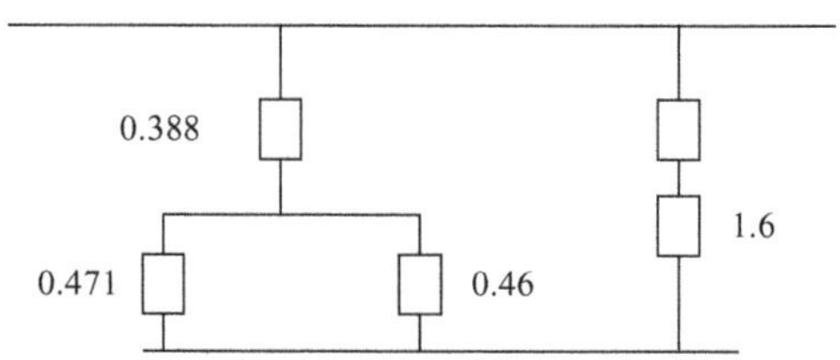

Figure 3.18 Further simplification

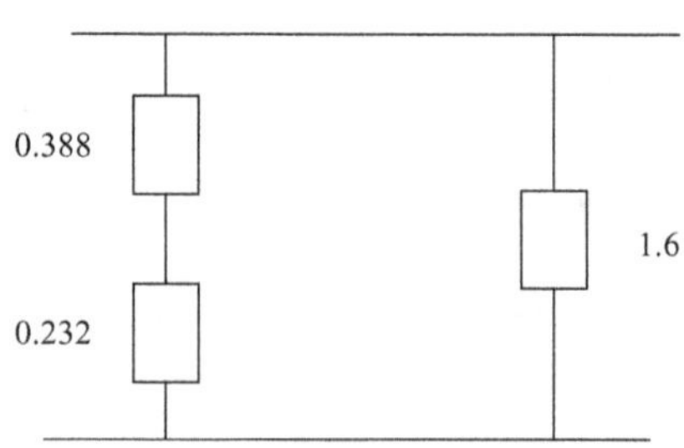

Figure 3.19 Simplification to give parallel impedances

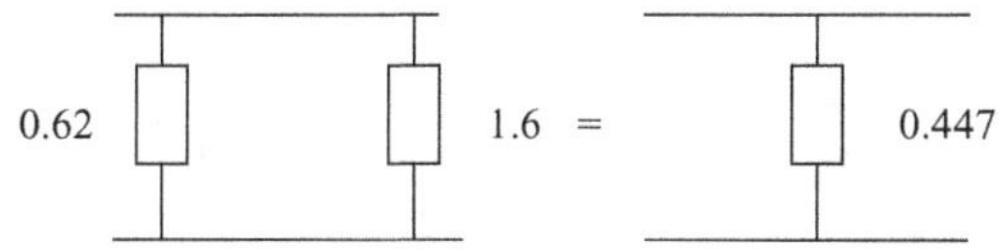

Figure 3.20 Final simplification

The total pu reactance in the circuit is, therefore, 0.447 pu.
The fault MVA is derived from

$$\frac{\text{MVA}_{\text{base}}}{\text{Xpu}_{\text{base}}} = \frac{100\,\text{MVA}}{0.447}$$

$$= 223.7\,\text{MVA}$$

Therefore:

$$\text{three-phase fault current} = \frac{223.7}{\sqrt{3} \times 11}.$$

$$= 11.74\,\text{kA rms.}$$

The following are further problems for you to resolve in your own time, using the previous complex network example.

3.2 Problems

(1) Calculate the symmetrical fault current that would have flowed if circuit breaker 'A' had been open.

(2) Similarly, calculate the symmetrical fault current that would have flowed if circuit breaker 'B' was the only circuit breaker open.

(3) What would have been the symmetrical fault current if both circuit breakers 'A' and 'B' had been open?

(4) Calculate the symmetrical fault current if the pu reactance of all of the transformers was 1 pu and all breakers were closed.

Chapter 4
Symmetrical and asymmetrical fault currents

In the previous chapter, we saw how to calculate the symmetrical fault current. This is important from the rms heating point of view, but for distribution switchgear engineers, the asymmetrical current is of much greater importance for a number of reasons. The rate of rise of current is higher under symmetrical fault conditions but the peak current of a fully asymmetrical fault current will induce the maximum electromagnetic force, and therefore stress, on conducting components. In addition, the total contact loading, which is the sum of electromagnetic and spring loading, must be sufficient to prevent contact burning. A fully asymmetrical current, as it is offset, will consist of major and minor loops. The time between current zeros in a major loop will therefore be greater than that implied by the power frequency of the system. This will induce greater stress on the interrupting system being used and, therefore, must be proven by test.

Medium- and high-voltage transmission and distribution of electricity takes place within a three-phase system. The voltage of each phase being displaced from the others by 120 electrical degrees. This means that the initiation of a three-phase fault will always take place with a finite value of prospective short-circuit current in at least two phases. In practice, the probability will be that all three phases will have some measure of finite prospective current. However, in an inductive circuit, an instantaneous value of prospective current cannot flow as the back e.m.f. of the circuit will provide an equal and opposite prospective current, starting the actual current flow at zero and giving the subsequent current wave a measure of asymmetry. An example showing the maximum phase values of asymmetrical current and d.c. component of transient current is given in Figure 4.1.

In the example shown in Figure 4.1, the maximum current asymmetry occurs in the centre, or Y, phase. An analysis of the total asymmetrical current flowing shows that it is made up from two components: (i) a prospective power frequency current and (ii) a decaying transient d.c. current. In other words, the instantaneous value of total current at any time after fault initiation can be derived from the expression:

$$i = [\text{steady state current} + \text{transient current}].$$

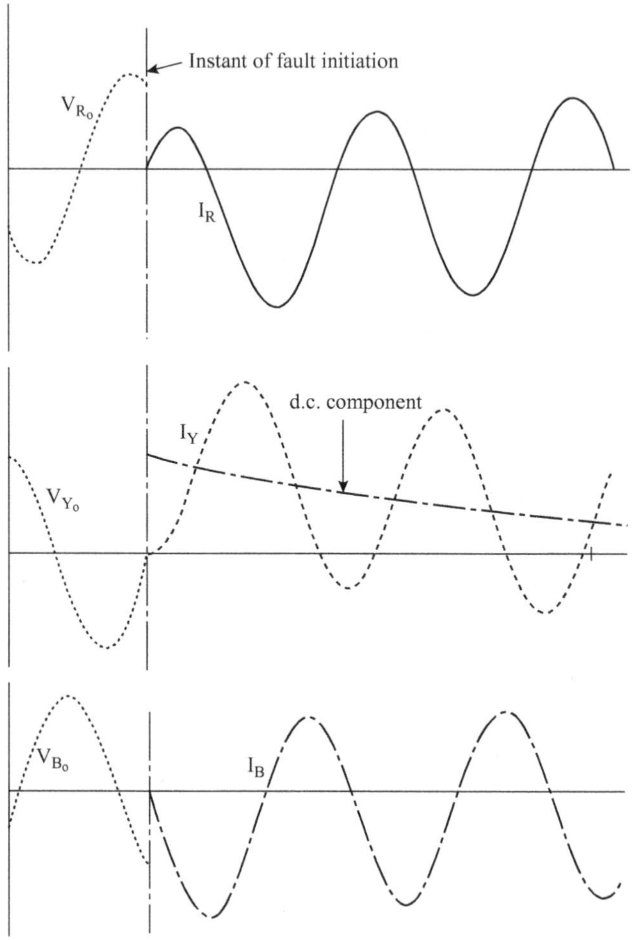

Figure 4.1 Oscillogram of a three-phase asymmetrical fault

Given that i = instantaneous value of current, I = the symmetrical peak current, ϖ = radians/s, ϕ = switching angle, t = time after fault initiation, L = circuit inductance, R = circuit resistance, the steady state current at any instant in time, i', is given by the expression

$$i' = I \sin(\varpi t + \phi)$$

and the value of the transient d.c. component of current, i'', at any instant in time after fault initiation is given by the expression

$$i'' = I \sin \phi e^{-Rt/L}.$$

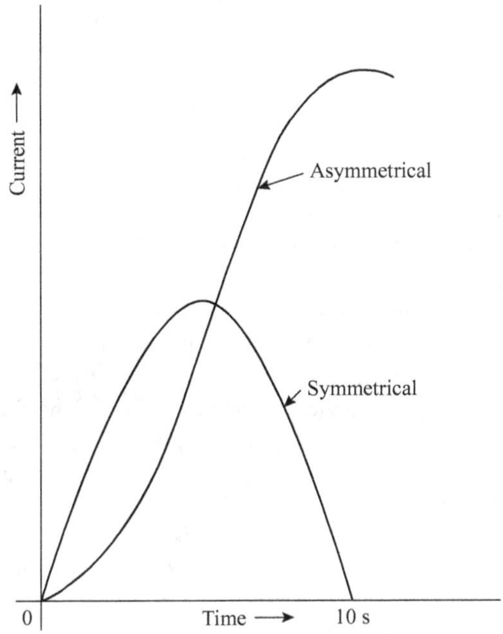

Figure 4.2 Extremes of initial current – symmetrical and asymmetrical

It follows then that the total instantaneous value of current for a given time after fault initiation, i, is given by the expression

$$i = [I \sin(\varpi t + \phi) - I \sin \phi e^{-Rt/L}].$$

The two extremes of initial fault current that can flow, symmetrical and asymmetrical, are shown in Figure 4.2.

The previous figure adequately demonstrates three areas of importance to the plant engineer. These are the rate of rise of current, the peak current and the time between current zeros. It will be seen that the symmetrical current waveform provides a very much higher rate of rise of current and will, therefore, have a significant influence on the making capacity of a circuit breaker. In this respect, the symmetrical current waveform is much more onerous than that of the asymmetrical current. Conversely, the peak of the asymmetrical current can be seen to be considerably higher than that of the symmetrical current, requiring much higher contact loading to prevent burning. As circuit breakers generally require a naturally occurring current zero to extinguish fault current, the time to current zero of the asymmetrical current may present the circuit breaker with some difficulty as the i^2t, or let through energy that has to be controlled by the interrupter, is very much greater. However, this difference in times to current zero will diminish with time as the d.c. component decays. It will, therefore, be a function of the time from fault initiation to initiation of arcing within the arc control device of the circuit breaker.

4.1 The rate of decay of the d.c. component

The rate of decay of the d.c. component is important to the switchgear engineer as it has a significant influence on the peak current and the time to current zero. These affect the contact loading, electromagnetic forces and the arc energy that has to be safely handled. In the earlier expression for the instantaneous value of current, the second element within the brackets gives the value of d.c. current. For a given point on wave of fault initiation, time after fault and symmetrical current, the value of the power factor has a large influence on rate of decay of the fault current. This is demonstrated in the following graph (Figure 4.3), which plots the d.c. component of current against time for power factors of 0.05, 0.1 and shows that the rate of decay increases with power factor.

The actual value of d.c. component to be used for type tests is specified in IEC standards. This is in the form of a graph which plots the d.c. component against the time interval from initiation of the short circuit. Currently, work is being carried out to provide two curves for distribution switchgear, one for standard applications having a time constant of 45 ms, and a new one for special applications that involve high d.c. components, with a time constant of 120 ms. These curves are shown in Figure 4.4.

For relay-operated circuit breakers, the required percentage d.c. component is determined by adding a half-power frequency cycle to the opening time of the circuit breaker, to represent the relay operating time. The total time then indicates the percentage d.c. component. For a self-tripping circuit breaker, no time is added to the opening time.

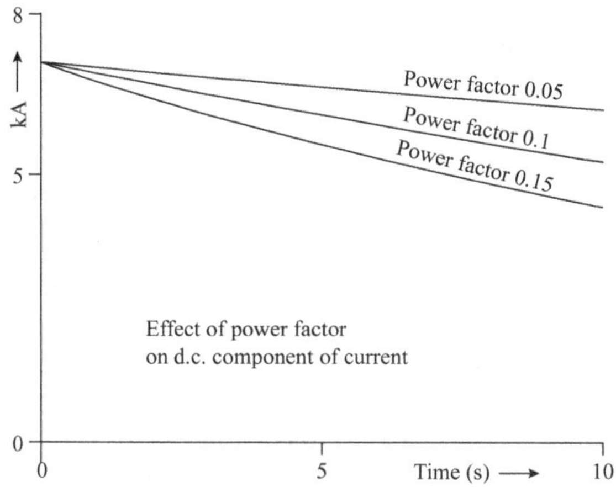

Figure 4.3 Relationship between power factor and d.c. component

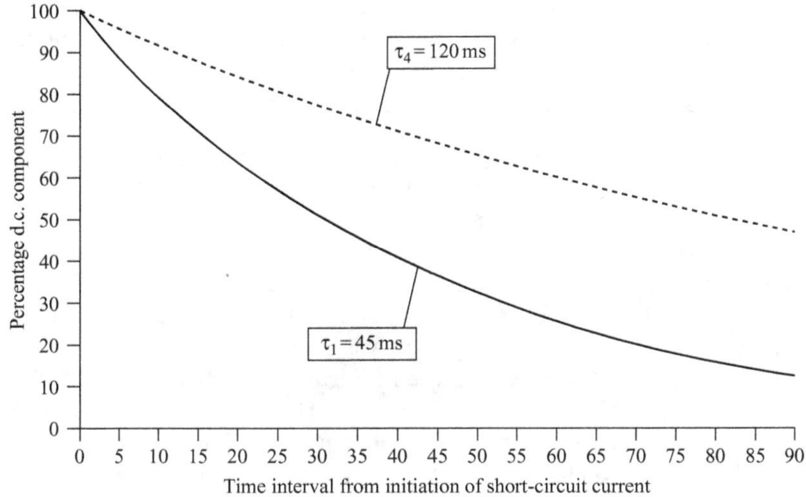

Figure 4.4 *Relationship between percentage d.c. component and the opening time of the circuit breaker*

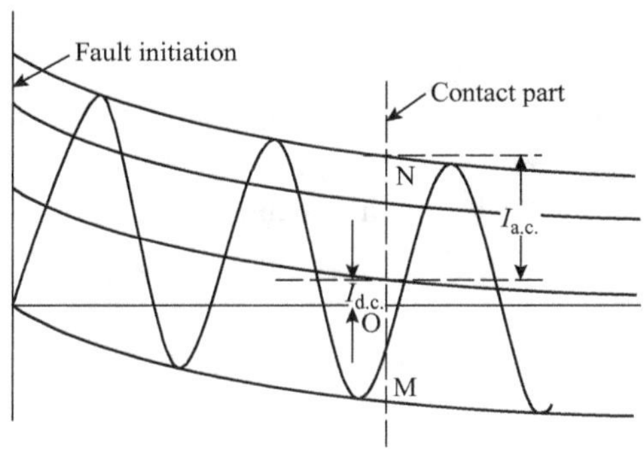

Figure 4.5 *Graphical representation of the d.c. component*

Referring to Figure 4.5, mathematically, the percentage d.c. component is given by

$$\frac{\text{ON} - \text{OM}}{\text{MN}} \times 100$$

Example 4.1

A relay operated circuit breaker having a minimum opening time of 35 ms is to be installed in a 50 Hz circuit. If the duty is one of normal power distribution, what is the percentage d.c. component that must be proven?

As a half-power frequency cycle in a 50 Hz circuit is 10 ms, the total time to be used in the graph will be 10 ms + 35 ms = 45 ms. And the corresponding percentage d.c. component will therefore be 37 per cent.

The high d.c. applications requiring the 120 ms time constant curve are usually associated with locations near to generators and free rotating plant, such as offshore oil production platforms. Until the introduction of the second, high d.c. component curve, manufacturers were offering either standard circuit breakers fitted with delayed trip mechanisms, or limited proven high d.c. component test evidence. In the event of a short circuit involving circuit breakers with a delayed trip, that trip would delay operation until the projected d.c. component was at the level proven during type tests. This was clearly a poor second choice when compared with a circuit breaker that was proven for this higher d.c. level of performance. This second curve will mean that purchasers may specify the new d.c. component in their call for tenders. However, if the demand for such switchgear is small, it may not justify the cost of redesign and proving tests, and other methods of dealing with the high d.c. component may have to be employed.

4.2 Decrement factor

A convenient way to calculate the decay of the d.c. component in a circuit is to use the decrement factor. The time constant of the circuit considered is R/L and, by definition, the decrement factor is given by $e^{-R/L\times 0.01}$.

This means that at $t = 0.01$ s, the instantaneous value of the d.c. component of current

$$i'' = I \times \mathrm{df}$$

where df is the decrement factor.

At $t = 0.02$ s,

$$\begin{aligned} i'' &= I \times e^{(-R/L\times 0.01)^2} \\ &= I \times \mathrm{df}. \end{aligned}$$

And at $t = 0.03$ s,

$$i'' = I \times e^{(-R/L\times 0.01)^3}$$

and so on. A numerical example will demonstrate how this works out in practice.

Example 4.2
If the d.c. component of current at the instant of short circuit is 1000 A, and the decrement factor is 0.8, calculate the curve of the d.c. component current for the first 5 ms of the fault.

Time after fault initiation	Calculation	The d.c. component of current
0	–	1000
0.01	1000×0.8	800
0.02	800×0.8	640
0.03	640×0.8	512
0.04	512×0.8	409.6
0.05	409.6×0.8	327.68

The following numerical example demonstrates how circuit parameters and point-on-wave switching information can be used to determine values associated with the resulting short-circuit current.

Example 4.3
Determine the value of the peak current and the time to the first current zero for a single-phase short-circuit test where the circuit parameters have been set to produce a steady state short-circuit current of 5 kA, given that the frequency is 50 Hz, the power factor 0.1 and that the fault is initiated 45° after the start of the positive wave of the voltage.

A graphical representation of the conditions specified is shown in Figure 4.6.

The prospective symmetrical current is that current which would flow if there were no back e.m.f. generated within the circuit.

The peak of the prospective symmetrical current

$$\begin{aligned} I &= \sqrt{2} \times 5000, \\ &= 7071\ \text{A}. \end{aligned}$$

At the instant of switching, $t = 0$, the d.c. component,

$$\begin{aligned} i'' &= -I \sin\phi \\ &= -7071 \sin(-39.3^\circ) \\ &= 7071 \sin(39.3^\circ) \\ &= +4478.64\ \text{A}. \end{aligned}$$

The significance of the positive sign is that the d.c. component of current is positive.

This, of course, is because at the instant of switching, the prospective symmetrical current can be seen in Figure 4.6 to be negative.

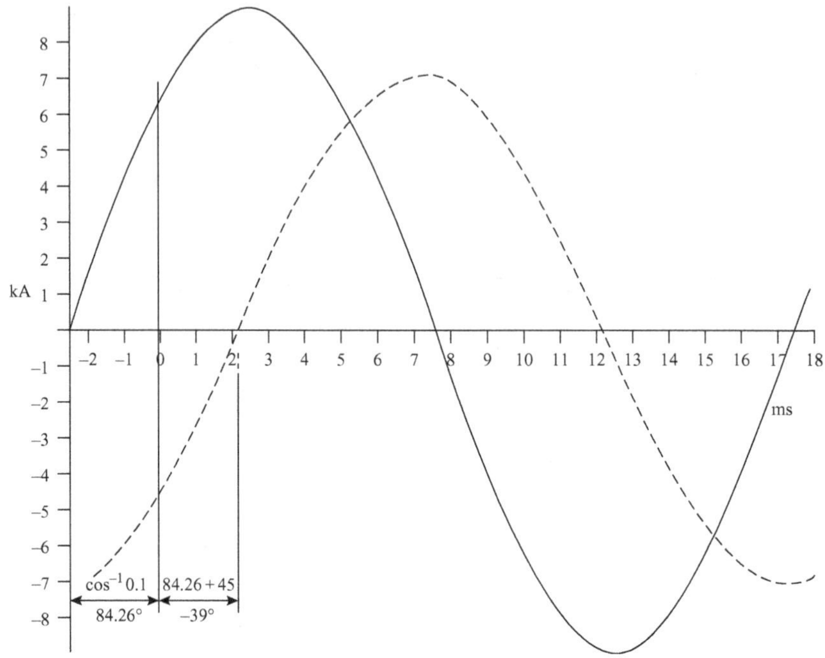

Figure 4.6 Graphical representation of voltage and prospective current

Given that the power factor, $\cos\phi = 0.1$, $\phi = 84.3°$ and $\tan\phi = 10 = \varpi L/R$,

$$L/R = 10/\varpi = 10/314.$$

Therefore

$$\text{the d.c. component of current} = 4478.64\mathrm{e}^{-31.4t}$$

and

$$\begin{aligned}\text{the decrement factor df} &= \mathrm{e}^{-R/10.001}\\ &= \mathrm{e}^{-0.0314}\end{aligned}$$

and

$$\begin{aligned}\log_{10}\text{df} &= -0.0314\log_{10}\mathrm{e}\\ &= -0.01365.\end{aligned}$$

Therefore

$$\text{df} = 0.9699.$$

The calculated values of the d.c. component of current, the symmetrical current and the resulting asymmetrical current over the first 18 ms of current flow will be as given in the following table:

Time from fault initiation (t)	Prospective symmetrical current (i')	The d.c. component of current (i'')	Asymmetrical current $i' + i''$
0	−4478.64	4478.64	0
0.001	−2534	4344	1810
0.002	−372	4213	3841
0.003	1826	4086	5912
0.004	3848	3963	7811
0.005	5491	3944	9435
0.006	6599	3728	10327
0.007	7053	3616	10669
0.008	6832	3507	10339
0.009	5936	3402	9338
0.01	4456	3299	7755
0.011	2534	3200	5734
0.012	385	3104	3489
0.013	−1816	3010	1194
0.014	−3837	2920	−917
0.015	−5484	2832	−2652
0.016	−6595	2746	−3849
0.017	−7060	2664	−4396
0.018	−6836	2584	−4252

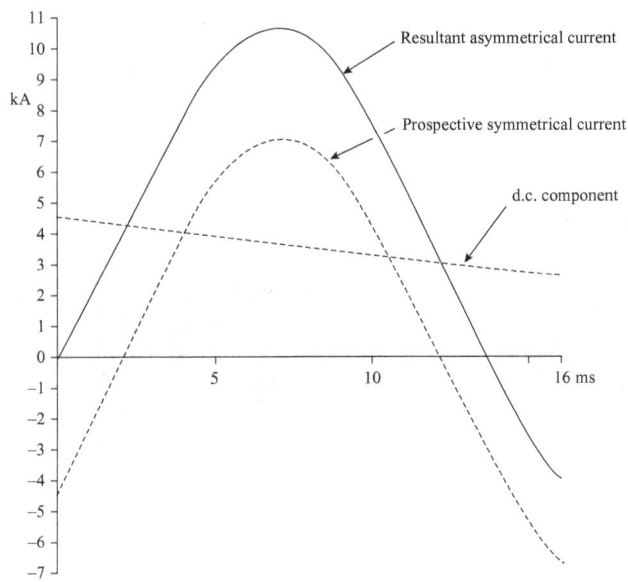

Figure 4.7 Graphical representation of the currents

The above results are shown graphically in Figure 4.7. It will be seen that the peak current was 10,699 A and that the time to the first current zero was 13.6 ms.

The worst case condition, in terms of peak current magnitude, would only occur if the fault had been initiated at the point in time when the prospective symmetrical current was passing through a maximum value. This is demonstrated in the following example:

Example 4.4
Determine the maximum peak current and time to current zero using the parameters given in the previous example assuming that the fault is initiated as the prospective symmetrical current wave is passing through a maximum value.

As before, the symmetrical peak current $i' = \sqrt{2} \times 5000 = 7071$ A.

Adopting the same calculation method as demonstrated in the previous example, the following table can be constructed:

Time from fault initiation (t) (ms)	Prospective symmetrical current (i')	The d.c. component of current (i'')	Asymmetrical current ($i' + i''$)
0	−7071	7071	0
2	−5720	6638	918
4	−2184	6232	4048
6	2079	5851	7930
8	5720	5493	11213
10	7071	5157	12228
12	5720	4841	10561
14	2184	4548	6732
16	−2079	4267	2188
18	−5720	4005	−1715
20	−7071	3760	−3311

This is shown graphically, in Figure 4.8.

It will be seen that the asymmetrical current does not reach a natural current zero until some 16.8 ms after the fault was initiated and the peak current experienced was some 12,228 A. The extended time to the first current zero, and the magnitude of the current peak would both have a significant influence over the design of switchgear in terms of mechanical stress, heat generation and contact loading.

4.3 Problems

(1) A single-phase 1500 MVA short-circuit test is to be carried out on a 33 kV normal distribution circuit breaker. If the power factor of the fault is 0.1 and the fault is

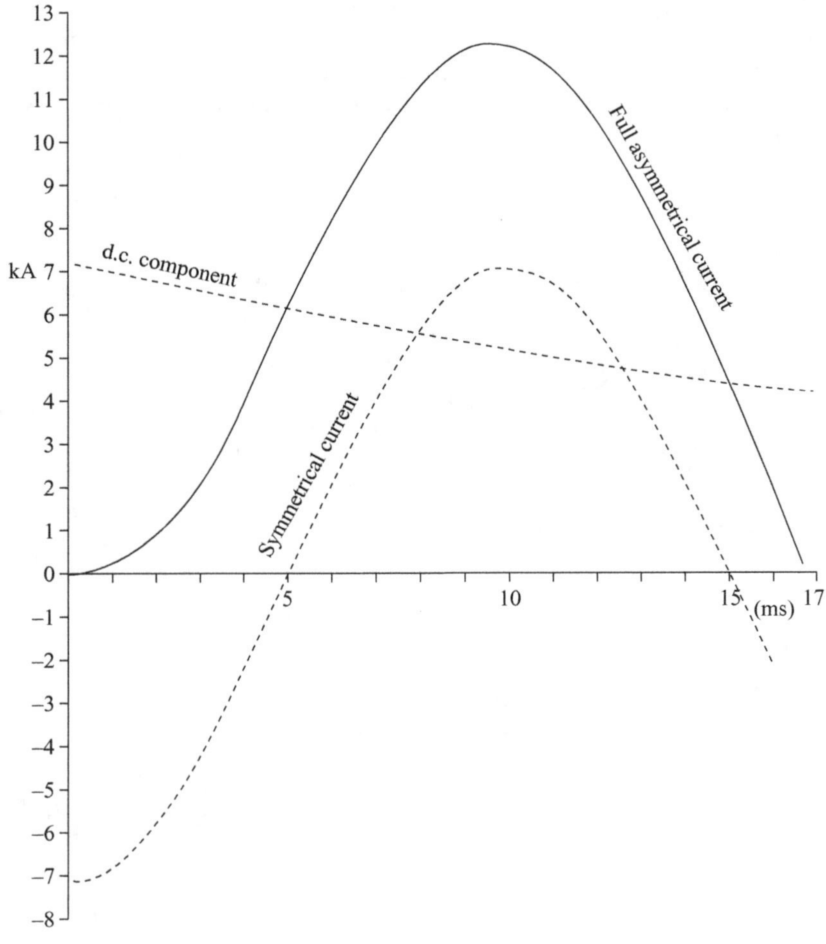

Figure 4.8 Graph of the full asymmetrical condition

initiated 75° after the start of the voltage cycle, determine the maximum value and duration of the first peak current.

(2) Using the circuit breaker and parameters given in Problem (1), determine the maximum value and duration of the first peak current if the point-on-wave of fault initiation coincided with current zero of the prospective symmetrical current.

(3) If the circuit breaker in Problem (1) had an opening time of 40 ms, was relay operated, and was to be located close to generators, or large capacity free rotating plant, determine the maximum peak current and arc duration assuming that the point-on-wave of fault initiation coincided with current zero of the prospective symmetrical current.

(4) An 11 kV relay-operated circuit breaker, having an opening time of 0.04 ms is to be located at a point in a system having a short-circuit fault level and power factor of 350 MVA and 0.07, respectively. Determine the maximum peak current against which the circuit breaker will have to close and the d.c. component which it will have to accommodate.

Chapter 5

Electromagnetic forces and contact design

It is a fundamental phenomenon of electricity that a force is exerted between conductors carrying electrical current. Under normal load current conditions, these forces are very small, however, many engineers will not be aware of the enormous forces that are generated when the normal current is replaced by a short-circuit current which can be 40–50 times larger in magnitude. This force generated phenomenon forms the basis of many desirable aspects of electrical engineering, such as the operation of measuring instruments and electrical motors, but in switchgear these forces are potentially dangerous in terms of the stresses induced in both the conductors and their supporting insulators.

Two factors influence the magnitude of the electromagnetic force that will be experienced. These are the strength of the magnetic field and the current flowing. The field strength can be derived from Laplace's Law. This states that the field strength created at a point in space due to the passage of electric current through a conductor is inversely proportional to the square of the distance between that point and the conductor, and is directly proportional to all other factors. The force experienced by a conductor in a magnetic field is derived from Biot–Savart's law in that the force is proportional to the flux density and the length of the conductor.

Figure 5.1 shows two parallel conductors, length L (m) and spacing D (m), one carrying the return current of the other in a circuit. It will be seen that the conductors experience a repulsive force F(N) due to the magnetic flux produced by this current.

The magnitude of this force is given by the expression:

$$F = \frac{0.2 \times L \times I^2}{D} \tag{5.1}$$

where F is the force in Newtons, L the length of conductor in metres, D the conductor spacing in metres and I the instantaneous value of current in kA.

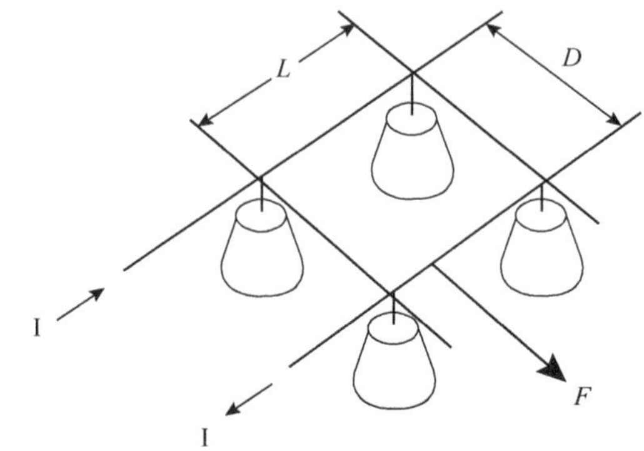

Figure 5.1 The force on parallel conductors carrying current

Example 5.1

Calculate the force acting on each of two parallel conductors having a length of 1 m and a separation of 20 cm given that the peak current flowing is 10 kA.

Force on each conductor

$$F = \frac{0.2 \times 1 \times 10^2}{0.2} = 100\,\text{N}$$

Assume that the arrangement of the busbars and the spacing of their supports are the same as that shown in Figure 5.1. It will be seen that the passage of a peak fault current of 10 kA will impose a point load at the upper end of each support of 50 N.

If the fault current had been 40 kA, four times the original value, the point loading would have been 800 N as the force is a function of the current squared. These peak forces also impose a bending moment on all support insulators which are not in line with the direction of the force.

The calculation of the mechanical stress in support insulators is, therefore, of paramount importance at the design stage.

It will be seen that Figure 5.1 also indicates the direction of the force. This is also of vital importance to the switchgear designer. The direction of the electromagnetic force can be simply determined by examination of the lines of flux produced by the passage of current. Figure 5.2 shows current passing through a conductor and the direction of the resulting magnetic flux. Obviously, current in the reverse direction would produce flux in the reverse direction.

We now examine the flux produced by parallel, same direction, currents and that produced by parallel, opposite direction, currents. Figure 5.3 graphically shows that in one condition we have the flux density augmented between the conductors and in the other it is diminished. A dot in the centre of the conductor indicates that the current is entering perpendicular to the page, and a cross that it is leaving.

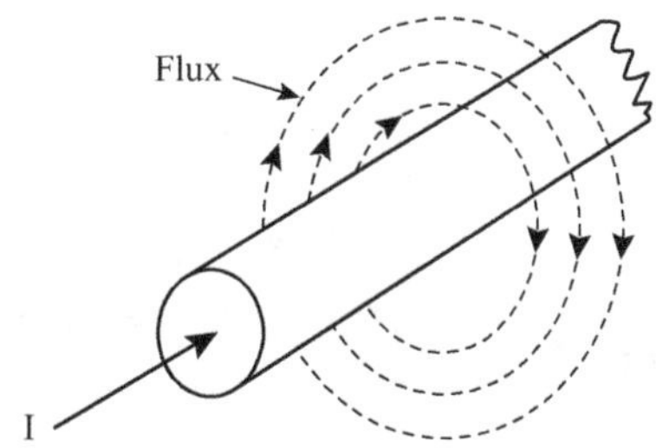

Figure 5.2 Direction of flux produced by a current in a conductor

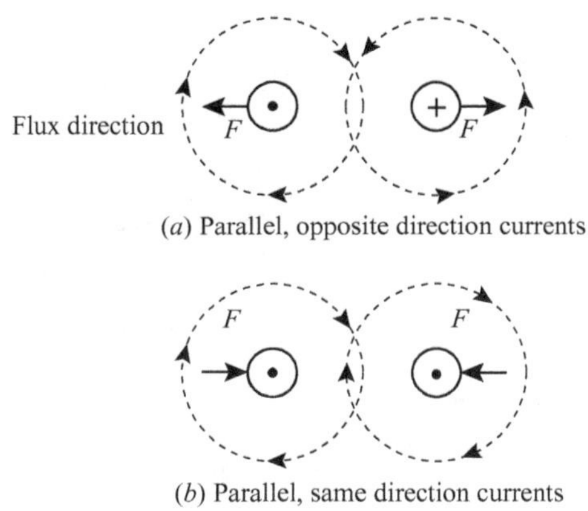

Figure 5.3 Diagram of flux and force directions in conductors

The best way to imagine the action of electromagnetic flux is to remember that lines of flux are always trying to maximise their length. They do this by applying a force on the conductor, or conductors, in an attempt to move them into a more favourable position for this maximisation to take place. It follows that a current carrying conductor which is not straight will experience bunching of the lines of force at the inside of the bend and the force produced will therefore tend to straighten the conductor, as illustrated in Figure 5.4.

This single conductor force is utilised in certain designs of air-break switchgear to assist the arc to move quickly in the direction of arc chutes for extinction, as shown in Figure 5.5.

In Figure 5.5, the slot in the moving contact causes the fault current to move down the moving contact and then upwards again. This will cause it to meet the arc root at about 90° and, therefore, will apply a driving force to the arc in an upwards direction, towards the arc control device, the arc chute, which is not shown.

Knowledge of the direction of the force generated within a single phase assists the switchgear designers in that the force can be used to augment the operation of circuit earthing devices and contact loading.

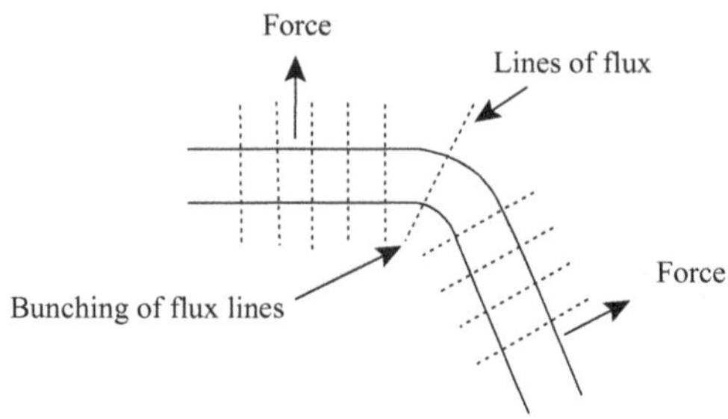

Figure 5.4 Force produced in a conductor by the passage of current

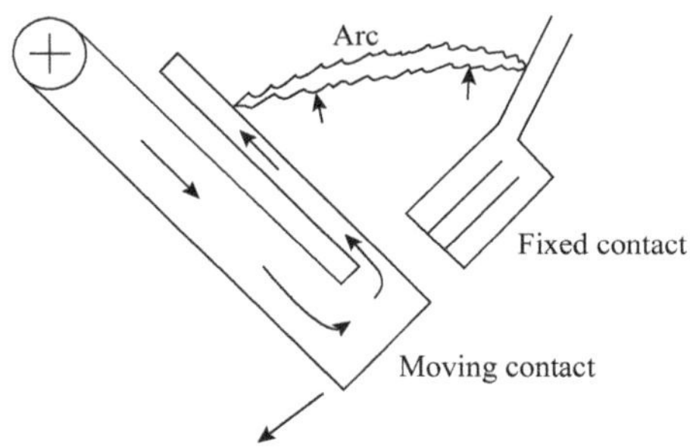

Figure 5.5 Electromagnetic drive applied to an air-break circuit breaker

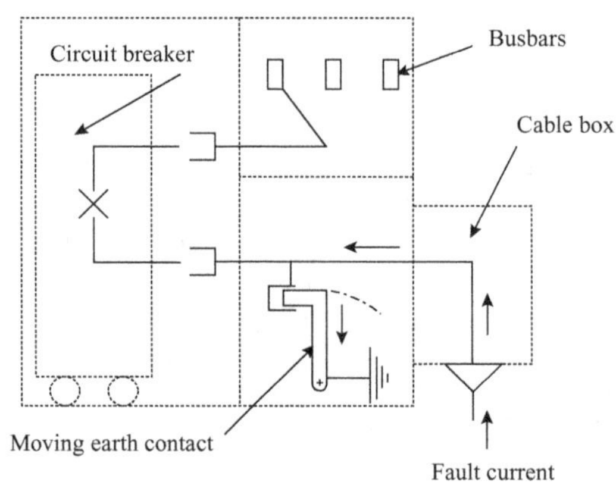

Figure 5.6 Typical architecture for a horizontally isolated circuit breaker

Let us look first at circuit earthing devices. With one or two exceptions, the typical architecture for a horizontally isolated circuit breaker requires a separate cable earth switch. This is to ensure that safety measures can be put in place while work is being carried out on the system. Such an architecture is shown in Figure 5.6, which shows the relationship between the circuit breaker, the busbar chamber and the cable chamber.

It will be seen that the moving contact of the circuit earthing device will experience a driving force towards the closed position from the fault current, as soon as the current starts to flow. If the moving contact had been arranged to close in a clockwise direction, there would have been a tendency for the moving contact to be forced towards the open position. To overcome this, the operating mechanism for the switch in this arrangement would need to provide much higher forces in order to close successfully against a full short-circuit fault.

Switchgear designers can also take advantage of the forces produced by parallel, same direction, currents in conductors. This can be seen in the design of isolating contacts, similar to those shown on the circuit breaker in Figure 5.6. The design of isolating contacts ranges from simple two-contact finger arrangements to complex multi-finger assemblies. Consider a simple two-contact finger isolating contact arrangement, shown diagramatically in Figure 5.7.

It will be appreciated from Figure 5.3 that the electromagnetic forces produced by the passage of current will cause the two contact fingers to be attracted to each other, increasing the loading at the point of contact. As the current carrying ability is a function of contact loading, these electromagnetic forces will increase the fault current that can be carried safely. It will also be appreciated that the forces for a given fault current will be far greater on the isolating contacts than on the actual phases, as these forces are inversely proportional to the conductor spacing. See expression 5.1.

Example 5.2

If the two-finger isolating contact shown in Figure 5.7 had a contact length L of 60 mm, and a contact spacing D of 35 mm. Ignoring any contact spring or blow-off forces, what would be the maximum electromagnetic force between the contact surfaces if the fault current was 25 kA?

The maximum electromagnetic force will occur at the peak of fault current. The peak of the fault current of 25 kA will be

$$2.5 \times I_{sc} = 2.5 \times 25\,\text{kA} = 62.5\,\text{kA}.$$

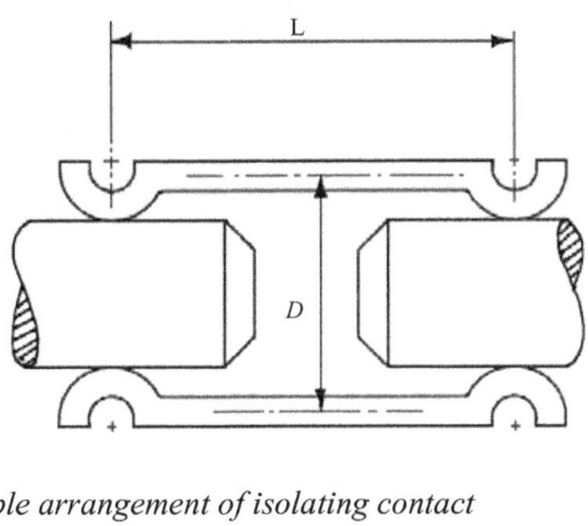

Figure 5.7 A simple arrangement of isolating contact

Assuming equal current sharing, the peak current seen by each contact finger will be

$$\frac{62.5}{2} = 31.25\,\text{kA}.$$

The force applied to the contact finger will be

$$F = \frac{0.2 \times L \times I^2}{D} = \frac{0.2 \times 0.06 \times 31.25^2}{0.035} = 334.82\,\text{N}.$$

Assuming one point of contact per contact end, the maximum electromagnetic contact loading force will be

$$= \frac{334.82}{2}$$

Electromagnetic contact loading will therefore be 167.41 N.

Example 5.3
If the same profile dimensions and fault current used in Example 5.2 were applied to a four-finger isolating contact, what would be the new electromagnetic force applied between the contact faces?

A diagrammatic end view of the contact assembly is shown in Figure 5.8. It will be seen that the four contact fingers in Figure 5.8 conform to Figure 5.3(b) in that they are parallel conductors carrying current in the same direction. As such, each finger will be attracted to each other. The forces of attraction being F_1, a force associated with the diametrically opposite contact, and F_2, the attractive force to each adjacent contact.

Assuming equal current sharing, the maximum peak current flowing in each finger will be $62.5/4 = 15.625$ kA and, as the contact spacing and length is identical to that used in Example 5.2, the force

$$F_1 = \frac{0.2 \times 0.06 \times 15.625^2}{0.035} = 83.7\,\text{N}.$$

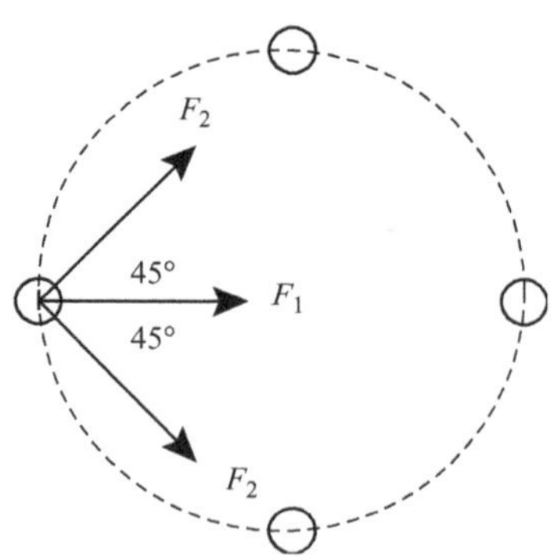

Figure 5.8 End view of the four-finger isolating contact assembly

The force between adjacent contacts F_2 will be greater than F_1 as the spacing between adjacent contacts is smaller. The actual spacing, from Pythagoras, being equal to

$$[(D/2)^2 + (D/2)^2]^{0.2} = [17.5^2 + 17.5^2]^{0.2} = 24.7\,\text{mm},$$

and, therefore, the force

$$F_2 = \frac{0.2 \times 0.06 \times 15.625^2}{0.0247} = 118.6\,\text{N}.$$

The force F_2 has a component which attracts each contact towards the contact surface. This, in the case of four equally spaced contacts is $F_2 \cos 45°$, which is numerically equal to: $118.6 \times 0.7071 = 83.86\,\text{N}$ per conductor and, as there are two adjacent conductors, the adjacent conductors will contribute to the contact loading force of: $2 \times 83.86 = 167.72\,\text{N}$.

The electromagnetic contact loading force for a fault level of 25 kA will, therefore, be:

$$\frac{167.72\,\text{N} + 118.6\,\text{N}}{2\ \text{(two ends to each finger)}} = 143.16\,\text{N}.$$

An example of multi-finger isolating contacts having a circular configuration can be seen in Figure 5.9 which shows two phases of a three-phase vacuum circuit breaker.

Because of other constraints, the switchgear designer cannot always use isolating contacts having a circular configuration, and has to use isolating contacts fixed to flat rectangular conductors, as shown in Figure 5.10 which shows the arrangement used on a 2000 A horizontally isolated vacuum circuit breaker.

Figure 5.9 The isolating contacts of two phases of a three-phase vacuum circuit breaker (courtesy of ALSTOM T&D Ltd)

Figure 5.10 The isolating contacts of a 2000 A horizontally isolated vacuum circuit breaker (courtesy of ALSTOM T&D Ltd)

Clearly, a contact arrangement as shown in Figure 5.10 will not provide any electromagnetic nip-on forces from the adjacent contact fingers. Indeed, measures should be taken to ensure that the adjacent fingers will not bunch together. However, the mating contacts on the reverse face will contribute to the nip-on force.

5.1 Contact loading

The required contact loading to prevent burning due to the passage of a given level of fault current will depend upon the following:

(1) the contact materials;
(2) the number of contact fingers;
(3) the number of guaranteed points of contact per contact finger end;
(4) the contact dimensions, that is, parallel length and spacing and
(5) the 'blow-off' or repulsive force at the contact interface.

The total contact loading will be numerically equal to

The applied spring load + the electromagnetic load − 'the blow-off' force.

The total required contact loading per contact point for various contact materials is given in Figure 5.11.

The information given in Figure 5.11 was accumulated over many years and has been found to be reasonably accurate. However, it must be remembered that the total

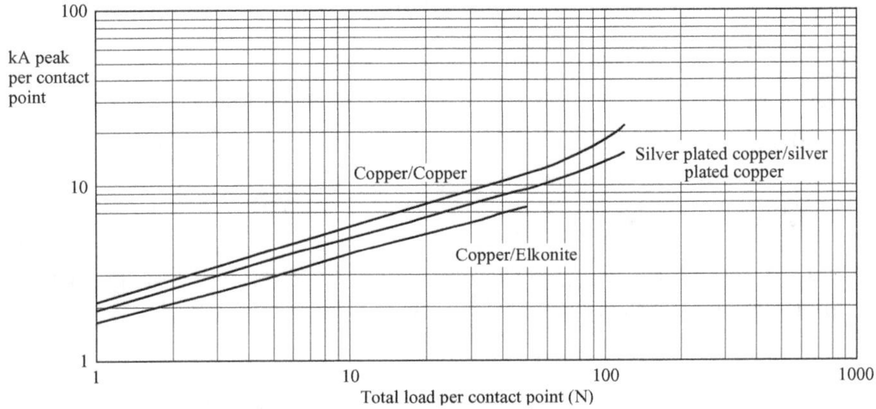

Figure 5.11 Total contact load required to prevent burning

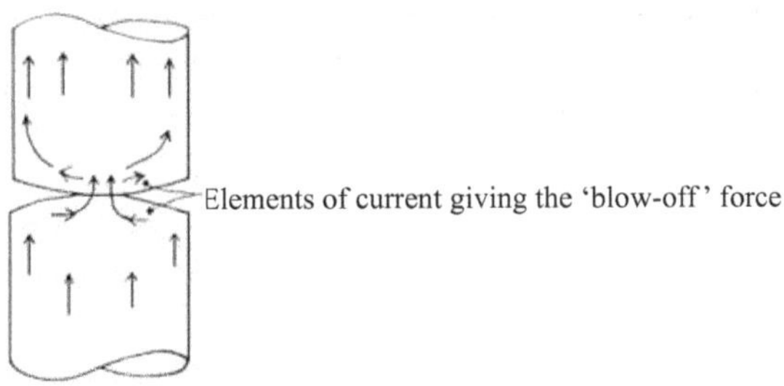

Figure 5.12 Contact interface and elements of current giving 'blow-off' current

load shown is per point of contact, which is the spring load plus electromagnetic 'nip-on' force, less the electromagnetic 'blow-off' force. This latter force is the repulsive force generated by parallel, opposite direction, currents, as described in Figure 5.3. The value of this blow-off force can only be estimated by examination of the contact design in the region of the contact interface to determine the mean current spacing and parallel length. A diagram illustrating this is given in Figure 5.12.

Example 5.4

Consider the application where a four-finger copper isolating contact having dimensions as given in Example 5.3 was to be used at a fault current rating of 25 kA. Given that the spacing and parallel sections of the currents in the region of the points of contact were estimated to be 2 and 5 mm, respectively, what spring loading per point of contact will be necessary to be able to withstand the passage of the fault current?

We have already calculated in Example 5.3 that the electromagnetic nip-on force from the fault current is 143.16 N per point of contact. We now need to determine the blow-off force from the estimated spacing and parallel lengths of current paths near to the points of contact.

The blow-off force per contact point will be:

$$= \frac{0.2 \times L \times I^2}{D}$$

$$= \frac{0.2 \times 0.05 \times 15.625^2}{0.02}$$

$$= 122\,\text{N}.$$

Therefore, the total force available electomagnetically will be the difference between the nip-on and blow-off forces

$$= 143.16\,\text{N} - 122\,\text{N}$$

$$= 21.16\,\text{N}.$$

However, Figure 5.11 shows that for copper contacts, a peak current of 15.625 kA will require a total external force per contact point of 85 N.

The contact springs will, therefore, have to supply the difference in loading between the electromagnetic force and the required loading from Figure 5.11.

Spring loading per point of contact

$$= 85\,\text{N} - 21.16\,\text{N}$$

$$= 63.84\,\text{N}.$$

Care should be taken to ensure that sufficient contact loading is provided to cater for all currents up to the maximum rating. It has been known for contact systems to perform without a problem at maximum short-circuit rating and yet give rise to contact burning at lower than maximum fault currents. This problem occurs when the electromagnetic contribution to the contact loading is significant compared to that provided by springs. The contact loading situation should, therefore, be checked over the full range of fault current up to the rated maximum.

5.2 Electromagnetic forces in three-phase faults

When a fault occurs in a three-phase system, the forces acting upon all three conductors balance each other. In other words, the system is a closed one and no external force will be produced.

This can best be demonstrated by examining the forces on each conductor of a three-phase busbar system which is arranged to have all three busbars in horizontal alignment, as shown in Figure 5.13.

The following can be seen in Figure 5.13:

(1) The direction of the forces in the outer phases is largely outwards.

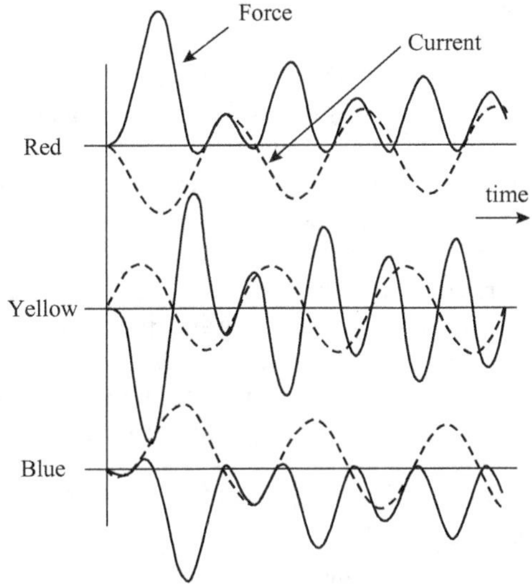

Figure 5.13 Forces in a horizontal three-phase system

(2) The direction of the forces in the centre phase is, more or less, equal in both directions.
(3) The frequency of the forces is twice that of the power frequency.
(4) At any one instance in time the forces in each direction are equal and opposite.

It follows then that the switchgear designer should arrange for any insulation bracing the busbars to be in line with the direction of the expected electromagnetic forces whenever possible. It is also good practice to examine all insulation supports carefully after carrying out short-circuit type tests, and not just rely upon a power frequency withstand test on the insulation. The rate of strain imposed upon the insulation is very high, typically reaching a peak in 5 ms, and as such, a change in the modulus of elasticity of the material should be expected. In other words, a static loading test will not predict the behaviour of the material under the dynamic conditions that will exist when a short-circuit fault is experienced.

5.3 Arcing contact tips

Up to now, we have only considered static conductors and isolating contacts which are only separated under no-load conditions. Arcing contacts differ, in that they are designed to close and open while carrying electrical current. This current can be very large and there is potential for arcing damage unless special measures are taken in the design of the contact system. Within oil and SF_6 switchgear, arcing contacts are usually provided with copper–tungsten tips attached to the copper contacts. These tips, which are sometimes known by the generic name, Elkonite, are manufactured by allowing copper to melt and permeate into a compressed block of tungsten powder.

This results in copper within a matrix of tungsten, which gives much lower ablation of the contacts when subjected to an electrical arc. However, there are three important issues that need to be recognised when this material is used. The first is that the resistivity of copper–tungsten is considerably higher than that of copper, requiring a much higher contact force than that for copper alone. This can be seen in Figure 5.11. The second is that copper–tungsten is a brittle material and, because the arcing contacts take the impact loads associated with a closing operation, it is better attached to the copper contacts by riveting as well as brazing. Material attached to reduce arc erosion is useless if it does not remain in place during the life of the contacts.

A classical application of brazing, screwing and riveting can be seen in the moving contact finger of some oil circuit breakers, where the arc resistant material is in the form of a tapped ring, screwed and brazed on to the end of its associated copper contact finger. The designer clearly having learned some expensive lessons when previously the tip became detached.

Electron beam welding has also been used very successfully to attach the copper–tungsten contact tips, although this method can be expensive. To electron beam weld, a copper layer several millimetres thick is left on the surface of the copper–tungsten block for welding purposes.

The final point to note is that copper–tungsten is a very expensive material and, therefore, should be used sparingly.

5.4 Contact entry profiles

A common mistake made by inexperienced switchgear designers is to provide both the fixed and moving contacts with a radiused entry profile. The danger associated with a radiused profile is that any lateral displacement will result in a wide variation in the contact entry forces. This is because of the angle of contact, as illustrated in Figure 5.14, where the moving contact is shown in three different alignments with

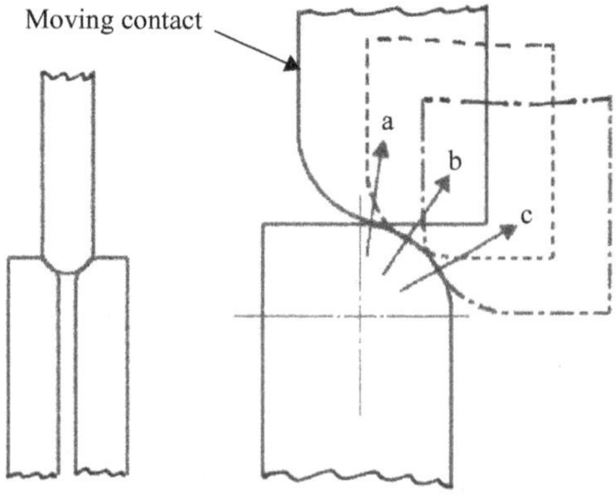

Figure 5.14 The influence of contact entry profile

the fixed contact. Lateral displacement can be the result of assembly setting errors or interface electromagnetic forces.

It can be clearly seen that the arrows a, b and c, indicating the normal to the contact face, have a wide angular variation which gives rise to wide differences in entry force.

In order to maintain a consistent entry force, a chamfer angle should be selected with an entry mouth greater than the maximum variation that the design allows.

5.5 Pre-arcing and contact burning

When contacts are moving towards each other, a point will be reached where the dielectric strength of the remaining contact gap is unable to withstand the voltage stress that is being imposed upon it, and electrical current will start to flow. This will result in a measure of arc erosion of the contacts taking place. At the instant when the contacts initially touch, there will be no contact load, and further erosion of the contacts will take place. This contact erosion will continue until the minimum contact loads to prevent burning are established. The current will have started at zero, at the instant of initiation of pre-arcing, and will have risen as a sinusoidal wave regardless of the contact loading and any contact burning. It is, therefore, important that the speed of contact closure is as fast as possible, and that contact loading is established as early as possible. However, as with most things in engineering, a compromise needs to be made between minimising pre-arcing by minimising the time to close and provide full contact loading, on the one hand, and the life, size and cost of the circuit breaker on the other. It must be recognised that pre-arcing cannot be eliminated and, therefore, contact closure under short-circuit fault conditions will always lead to a measure of contact erosion.

5.6 Contact misalignment and fault making capacity

At the instant of closure, the rate at which current rises will be a function of the point on the prospective voltage wave that flow commences. The highest rate of increase is that associated with a symmetrical fault current, which has, relatively, the lowest current peak. Conversely, a fully asymmetrical fault current will have the highest peak current but the lowest rate of current rise. In a three-phase circuit breaker, it would be normal to expect a degree of asymmetry in, at least, two phases under short-circuit conditions. However, some contact misalignment can be expected, which will mean that the contacts in each phase will begin conducting at different times. If a circuit breaker with misaligned contacts is closed onto an insulated three-phase fault, as would be used for type tests, the short-circuit current will start to flow when the second phase becomes conducting. The worst case scenario would be when this point of closure produces the start of a symmetrical current wave, in that this will produce the highest initial rate of current rise. When the third phase becomes conductive, a phase shift will take place, which could cause the current in one of the already conducting phases to change to a full asymmetrical wave. This would mean that

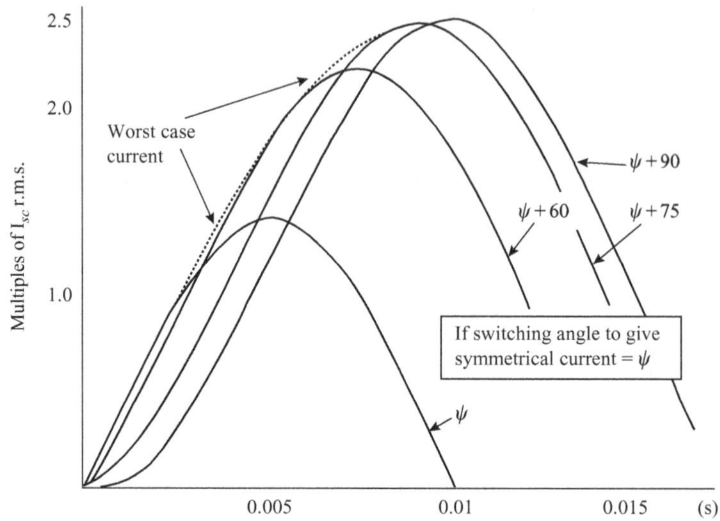

Figure 5.15 Worst case condition due to contact timing

the current in that phase would have started with the highest rate of current rise and then changed to reach the highest current peak. This then means that this phase would be carrying the highest current at any given time after the start of current flow; a very onerous condition for the contact system. It is, therefore, vital that the routine tests carried out on production circuit breakers have a contact closing timing spread which is not greater than that measured during the proving type tests. Figure 5.15 shows the worst case conditions, where the highest current in a given time after the start of current flow occurs due to a start which is initially symmetrical and subsequently changes to become a full asymmetrical wave.

5.7 Sliding frictional resistance of contacts

The sliding frictional resistance of engaged finger-type cluster contacts is an important factor in the design of switchgear. It is particularly important for equipment having a high short-circuit current rating, as both the number of contact fingers and their loading could be high. In the early days of isolatable vacuum and SF_6 switchgear for distribution applications, some designs, in many companies and countries, were a direct development of the then existing oil circuit breaker types. These designs retained many of the components and design features of oil circuit breakers, however, the designers failed to take into consideration the significance of the reduced weight that the new technology provided. Prototypes of the new switchgear very quickly showed up the consequences of this reduced weight as the circuit breakers could not be isolated, because there was insufficient weight to disengage the isolating contacts. In addition to the frictional resistance to movement within isolating contacts, a similar resistance is provided by the main interrupting contacts of both oil and SF_6 switchgear, which are of the sliding 'self-cleaning' type. Vacuum circuit breaker contacts are invariably of the 'butt' type and therefore do not offer any sliding frictional resistance.

The expression for frictional resistance is:

$$P = \mu F$$

where P is the frictional resistive force, μ the co-efficient of friction and F the force between the contact faces.

As Figure 5.11 indicates, a non-linear relationship is involved in the required contact load, and, from its trend, it follows that increasing the number of points of contact will reduce the total sliding frictional resistance. This is because the level of shared current per point of contact is reduced, and, therefore, the interface contact load required to prevent burning is also reduced. The following example illustrates the effectiveness of this method of reducing the sliding frictional contact load.

Example 5.5

Given that the short-circuit rating of a circuit breaker is 25 kA r.m.s. and assuming that the co-efficient of friction, μ, of plain copper contacts is 0.3, determine the total frictional sliding resistance of the six isolating contact assemblies of a circuit breaker, assuming one point of contact per end, for four, eight and 12 contacts per isolating contact cluster.

By definition, the peak current associated with a 25 kA r.m.s. rating is 62.5 kA peak. Therefore, with four points of contact per assembly, the peak current per point of contact is 15.625 kA peak.

From Figure 5.11, we can see that the load per contact point to prevent burning will be 350 N.

The total load per contact cluster will, therefore, be $4 \times 350 = 1400$ N
and, the total load per circuit breaker will be: $6 \times 1400 = 8400$ N
and the sliding frictional resistance will be: $0.3 \times 8400 = 2520$ N.

Similar calculations can be made for six and 12 contacts that will allow Table 5.1 to be constructed. This is graphically shown in Figure 5.16.

Table 5.1 Calculations based on Example 5.5

a No. of fingers	b 162.5/a kA/point	c Load/point (Figure 5.11)	$d = ac$ Total load per cluster	$e = 6d$ Total load per cb	$f = 0.3e$ Sliding frictional load per cb
4	15.625	350	1400	8400	2520
8	7.813	90	720	4320	1296
12	5.208	35	420	2520	756

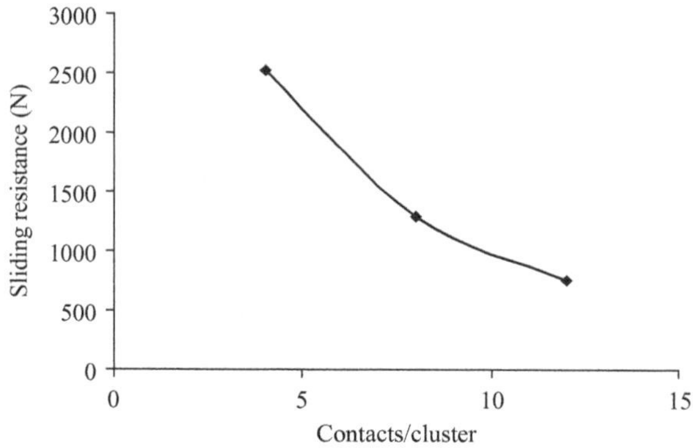

Figure 5.16 Reduction in sliding resistance with increase of contact fingers

Clearly, as can be seen above, the number of contacts within an assembly will have a marked influence upon the sliding frictional resistance.

5.8 Problems

(1) A circuit breaker has six isolating contact assemblies having 12 silver-plated contact fingers per phase. If electromagnetic forces acting upon the fingers in the event of a short-circuit can be ignored, what would be the maximum short-circuit rating of the isolating contact assemblies if there was one point of contact per finger end and the spring load per point was 200 N?

(2) If the fingers within the isolating contact assemblies given in Problem (1) were changed to plain copper at the contact faces, what would be the maximum short-circuit current that the assembly could safely carry?

(3) The isolating contact system of a circuit breaker consists of six copper fingers per pole having an effective length of 90 mm and a diametric spacing of 40 mm. Given that the spacing and parallel paths of current in the region of the points of contact were 1.5 and 4 mm, respectively, what would be the spring loading necessary at the points of contact to prevent burning, given that each finger had only one point of contact per end, and the short-circuit current was 40 kA r.m.s.

Chapter 6
Switching transients

A circuit breaker, when it closes or opens its contacts in an electrical circuit, causes energy stored within elements of the circuit to be redistributed over a very short period of time. During this period, voltages and currents can be produced which are far in excess of those which are normally present when the circuit is experiencing steady-state conditions. The levels of transient current and/or voltage produced during disturbance of an electrical circuit are of vital interest to those who design electrical systems because, without taking preventive or protective measures, damage to the circuit elements may take place.

There are three types of circuit element that make up an electrical circuit. These are the resistance R, capacitance C and the inductance L. These are normally distributed quantities within a circuit, and in most cases can be summed for calculation purposes.

Analysis of the three elements shows that two of them, the capacitance and inductance, store energy, while the third, resistance, dissipates energy. In an alternating current circuit, the stored energy within the capacitance and inductance is given by the expressions:

$\frac{1}{2}Li^2$ for inductive elements,

$\frac{1}{2}Cv^2$ for capacitive elements,

where i and v are the instantaneous values of current and voltage.

The third element within an electrical circuit is the resistance R, which dissipates energy. The value of the dissipated energy is given by the expression Ri^2.

In an alternating power system, the energy within the inductive and capacitive circuit elements is transposed between these elements as the instantaneous values of the current and voltage change. A sudden change in the circuit configuration, such as would be caused by the opening or closing of circuit breaker contacts, will cause the energy stored in the reactive elements to be redistributed. This redistribution cannot take place instantaneously, as the change in energy stored within the inductive reactance elements would require a change in current, and current cannot be created

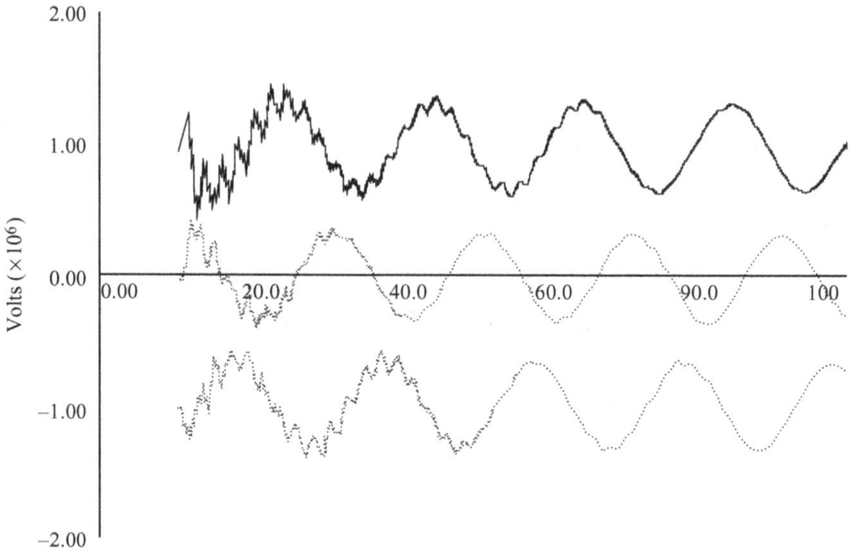

Figure 6.1 Result of a switching transient voltage study (courtesy of British Short Circuit Testing Station)

or destroyed instantaneously. Similarly, an instantaneous change in voltage stored within the capacitive reactance circuit elements would require the application of an infinitely large instantaneous current, which is not possible.

It can, therefore, be concluded that the redistribution of energy within a circuit following a disturbance will take a finite time to achieve. This time will be very short when compared with the normal power frequency of the circuit, and will depend upon the rate of energy storage and the rate of energy dissipation. Energy conservation must be observed in any calculation of transient duration.

In the case of complex, highly reactive circuits, the calculation of transient overvoltage generation by circuit breaker switching operations can be difficult to evaluate without resorting to transient analysing equipment. However, there are specialist companies that will carry out system studies that are able to show, with great accuracy, the abnormal conditions that could be expected. Figure 6.1 shows an example of the type of result that can be produced. In this case, the results of a study of switching overvoltages are shown. It will be noted that the transient voltage disturbance had a duration of about three cycles. This duration would have been reduced had the circuit contained more resistance.

6.1 The influence of system earthing on the transient recovery voltage

The type of earthing used on a distribution system will have a marked effect upon the transient recovery voltage that will appear across the contacts of a circuit breaker when

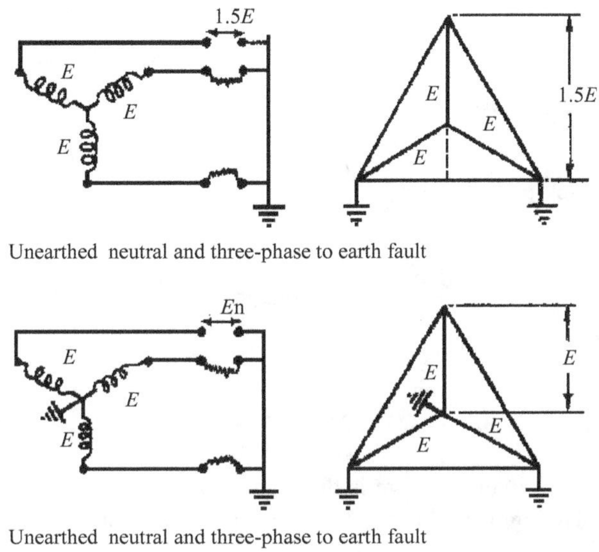

Figure 6.2 The influence of system earthing upon recovery voltage

it interrupts the flow of current. This is because within a three-phase system, current zeros will appear at a different time in each phase and the magnitude of the voltage which appears across the first phase to clear will be determined by whether or not the neutral of the system is earthed. The reason for this can be seen in Figure 6.2. This shows the voltage in each phase diagrammatically and vectorially. In the upper view, the system has an unearthed neutral and a three-phase fault to earth. When current is flowing in all three phases, the neutral point is held by capacitance at a central point in the vector diagram so that the voltage to earth of each phase is E. When the first phase clears the current, the neutral point moves immediately to a mid-point between the two conducting phases, giving a voltage across the first phase to clear of $1.5E$ and a reduced voltage across the conduction phases. In a system where the neutral of the system is effectively earthed, the neutral point is held rigidly at a central point in the vector diagram, as shown in the lower part of Figure 6.2, and each phase will see a recovery voltage with a value of E when it clears the current flowing.

We now consider the switching transients that are most likely to be experienced by distribution switchgear. The two characteristics of the transient recovery voltage (TRV) that most influence the performance of a circuit breaker when interrupting current are the rate of rise of recovery voltage (RRRV) and the TRV peak value.

6.2 The interruption of load current

When a circuit is carrying purely load current, the power factor of the circuit will be close to unity. This means that the system voltage and the load current are in phase with each other, with coinciding zero values. When at current zero, the load current

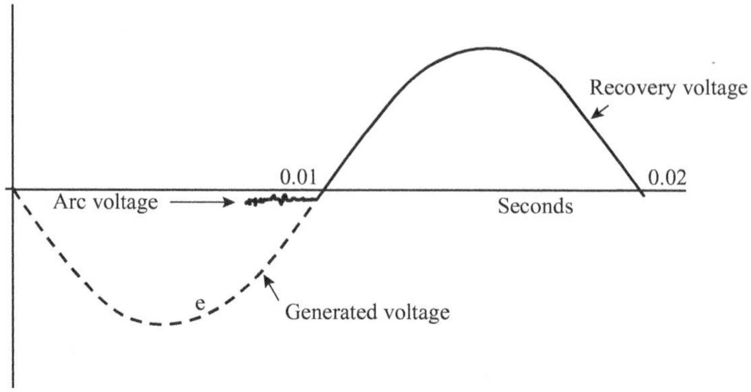

Figure 6.3 The interruption of load current

is interrupted, the recovery voltage is the relatively low-frequency system voltage. This is illustrated in Figure 6.3. In this figure, the arc suppression peak is very low and the relatively high circuit resistance will effectively damp out superimposed voltage oscillations.

6.3 The interruption of inductive current

A distribution circuit breaker must be able to interrupt highly inductive currents in two types of circuit: the first being the interruption of a short-circuit current and the second the interruption of the magnetising current of an unloaded transformer.

The transient recovery voltage of a highly inductive circuit is completely different from that associated with the interruption of normal load current. This is because of the different instantaneous values of system voltage at current zero. During the arcing period, the voltage across the contacts is equal to the small voltage drop of the arc column, but when the current is extinguished, the voltage across the contacts will rise to the system voltage. During the interruption of load current, the system voltage is virtually at zero when current ceases to flow and rises sinusoidally.

But, because of the lagging power factor, when current ceases to flow in an inductive circuit, the voltage on the system side contacts of the circuit breaker will attempt to rise to match the system voltage, which will be near its crest value. Due to the oscillatory nature of the recovery voltage, it will overswing the crest voltage and approach double its value. The natural frequency of a circuit containing capacitance and inductance is given by the expression:

$$\frac{1}{2\pi\sqrt{LC}}$$

This frequency will be superimposed upon the system voltage and the combination will form the recovery voltage which will appear across the contact gap when a circuit breaker clears an inductive fault. This is illustrated in Figure 6.4.

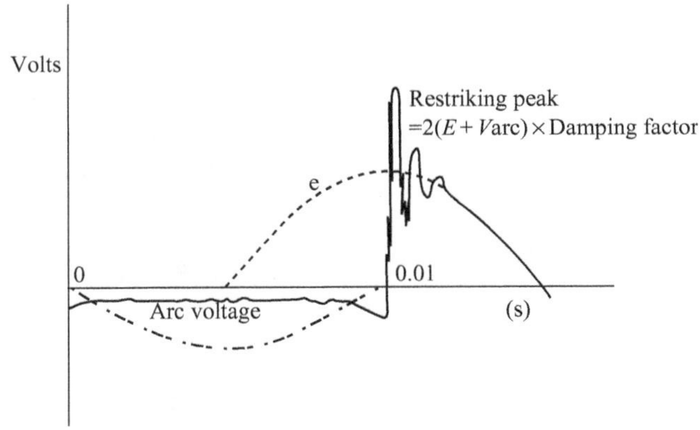

Figure 6.4 The transient recovery voltage associated with an inductive fault

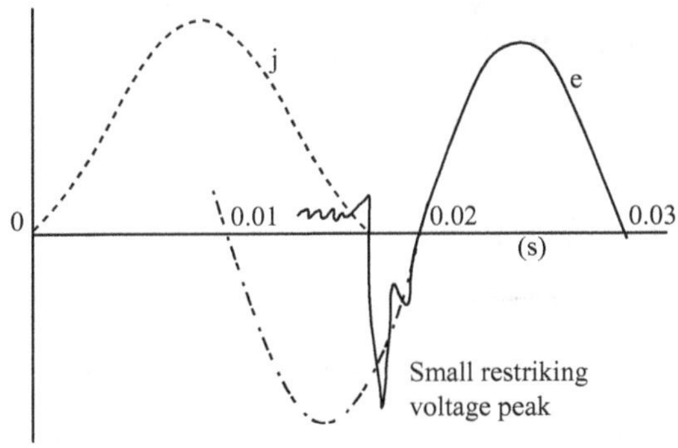

Figure 6.5 The influence of current asymmetry on TRV values

For an earthed neutral system, the peak voltage,

$$\mathrm{TRV}_{\mathrm{peak}} = 2 \times D_{\text{damping factor}}\left[E_{\text{system peak voltage}} + V_{\text{arc voltage}}\right],$$

and for an unearthed neutral system, the first phase to clear will see a voltage peak of:

$$\mathrm{TRV}_{\mathrm{peak}} = 1.5 \times 2 \times D_{\text{damping factor}}\left[E_{\text{system peak voltage}} + V_{\text{arc voltage}}\right].$$

If we now look at the TRV that will prevail under asymmetrical fault conditions.

Figure 6.5 shows the influence of current asymmetry upon the TRV. In a circuit that is highly inductive, the power factor will be low and lagging. This means that the current will lag the voltage by up to 90°. Under asymmetrical fault conditions, the time between current zeros of an asymmetrical major current will be far greater than the 10 ms that would normally be associated with a 50 Hz system. The following are some worked examples to help give a better understanding of the calculations involved in solving a relatively simple transient voltage generation problem.

Example 6.1

A three-phase alternator is connected to a busbar through a circuit breaker. The inductance of the alternator is 7 mH/phase and the capacitance to earth at the circuit breaker terminals is 0.006 μF. If a three-phase fault occurs at the circuit breaker terminals, determine the frequency of the restriking voltage and the time to peak value of the recovery voltage.

As was shown earlier, the natural frequency of the circuit is given by the expression:

$$\frac{1}{2\pi\sqrt{LC}}$$

Substituting the values given, the natural frequency will be:

$$\frac{1}{2\pi\sqrt{7\times 10^{-3}\times 6\times 10^{-9}}} = 24.6\,\text{kHz}$$

and, as the number of peaks is twice the frequency, the time to peak value will be:

$$\frac{1}{2f} = \frac{1}{2\times 24.6\times 10^{3}} = 20.32\,\mu\text{s}$$

Example 6.2

If the alternator and circuit breaker were rated at 11 kV, and the short-circuit fault location was as detailed in Example 6.1, calculate the recovery voltage across the first phase to clear and the rate of rise of recovery voltage, given that the system had an effectively earthed neutral. For the sake of this exercise, assume that there is no resistance in the circuit.

As the neutral of the alternator is effectively earthed, the voltage appearing across the first phase to clear will be:

$$\frac{11}{\sqrt{3}} = 6.351\,\text{kV}$$

and the associated peak system voltage will be:

$$\sqrt{2}\times 6.351 = 8.98\,\text{kV}$$

and, as there is no resistance in the circuit, the over-swing peak of the oscillatory component of voltage will be equal in value to the system voltage peak. Therefore, the peak value of the TRV will be:

$$2\times 8.98 = 17.96\,\text{kV}_{\text{peak}}.$$

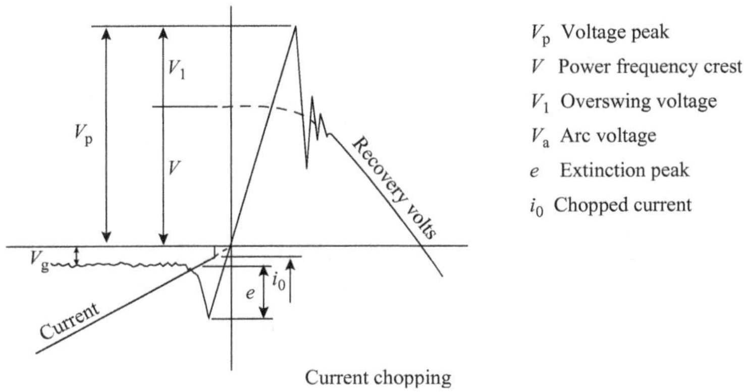

Figure 6.6 Voltages and current associated with current chopping

Finally, the rate of rise of the recovery voltage, the RRRV, will be:

$$\frac{\text{TRV}_{\text{peak}}}{\text{Time to peak}} = \frac{17.96\,\text{kV}}{20.32\,\mu\text{s}}$$

$$\text{RRRV} = 883\,\text{V}/\mu\text{s}.$$

6.4 The interruption of small inductive currents

When a powerful interrupter is called upon to interrupt a very small current, that current may be forced to a premature current zero. This is known as current chopping. When the small current being chopped is associated with a large value of inductance, such as the magnetising current of an unloaded transformer, the level of voltage generated may be large for a small value of chopped current. The voltages and current at the time of interruption are illustrated in Figure 6.6.

For a current, i_0, forced to zero, the value of voltage generated, e'', can be calculated by re-arranging the expression:

$$\tfrac{1}{2} \times L \times i_0 = \tfrac{1}{2} \times C[e'']^2$$

giving

$$e'' = i_0\sqrt{L/C}$$

In other words,

$$e'' = i_0 \times Z(\text{where } Z = \text{surge impedance of the transformer}).$$

Current chopping has been seen to occur with all types of interrupters, including oil, SF_6 puffer, SF_6 rotating arc and vacuum. Current chopping levels are often

Table 6.1 Current chopped by vacuum interrupters having different contact materials

Contact material	Current chopped
Chrome copper	1.6 A
Copper bismuth	5.0 A

Table 6.2 Surge impedance and magnetising current values for various transformer sizes

Transformer (kVA)	500	2000	5000
Magnetising current (a)	0.4	1.1	2.6
Surge impedance Z(kohms)	200	85	50

quoted by manufactures when their products display a relatively low level of current chopping, implying that greater levels of current chopping may produce dangerous overvoltages. However, this is very misleading. Consider, for instance, the case for vacuum interrupter contact materials, and in particular a comparison of chrome copper and copper bismuth contacts.

Table 6.1 shows that, during interruption, copper bismuth will chop more than three times the current that would be chopped by chrome copper under the same conditions. However, if we look at typical oil-filled distribution transformer magnetising currents and surge impedances we will find the values given in Table 6.2.

It is useful to note that, when calculating switching voltages, a damping factor (Df) of 0.25 can be used as this is a typical figure derived from core loss and measured values. The maximum possible voltage experienced by no-load switching, V_p, will be as shown in Figure 6.6:

$$V_p = E + e' + e''$$

With a damped oscillation about the system voltage E, the maximum voltage is calculated by:

$$V_p = \frac{V}{\sqrt{3}} + \mathrm{Df}\left\{\frac{V}{\sqrt{3}} + (Z \times i_0) + V_a\right\} \tag{6.1}$$

Example 6.3

An unloaded 11 kV, 500 kVA oil-filled transformer is switched by a vacuum circuit breaker having an arc voltage of 200 V. What is the maximum peak voltage that will be generated?

From Table 6.2, it will be seen that the magnetising current will be 0.4 A and its peak value will be $\sqrt{2} \times 0.4$ A, a value that will be chopped by the vacuum interrupter. Substituting the values into the expression (6.1):

$$V_p = \frac{11}{\sqrt{3}} + 0.25\left\{\frac{11}{\sqrt{3}} + (200 \times 0.4 \times \sqrt{2}) + 0.2\right\}$$

$$V_p = 26.352\,\text{kV}$$

It will be noted that the peak voltage calculated is well within the specified rated impulse level of 75 kV for 11 kV equipment and will not present an operational problem.

Similar calculations can be made for the other two transformer sizes given in the previous table. The maximum peak voltages to be experienced while switching off-load with either type of vacuum interrupter contact material are shown in Table 6.3.

It will be noted that the voltage generated when switching unloaded oil-filled distribution transformers with ratings up to 2000 kVA display the same value, regardless of the vacuum interrupter contact material, as both types of interrupter are capable of chopping the peak of the magnetising current. However, there is a difference in the value of voltage generated when switching a 5000 kVA transformer, because of the different current chopping levels displayed by the contact materials, but the peak value of 53 kV is still well within the rated impulse level of 11 kV equipment.

It could be argued that the current to be interrupted will increase once the transformer supplies a load and that this will increase the level of chopped current. However, this will not increase the voltage generated as the surge impedance of the transformer falls away rapidly when load currents start to be supplied.

Caution should be exercised when considering the use of vacuum interrupters to control dry-type transformers. This is because the surge impedances of these transformers are higher that those of the oil-filled type. For these applications, surge arrestors may be necessary to prevent possible insulation damage when these transformers are being switched off-load.

Table 6.3 Maximum voltage generation when off-load switching of transformers

Transformer (kVA)	500	2000	5000
Chrome copper	26.352 kV_{peak}	41.04 kV_{peak}	27.96 kV_{peak}
Copper bismuth	26.352 kV_{peak}	41.04 kV_{peak}	53.0 kV_{peak}

6.5 Capacitor switching

In addition to the more obvious applications such as single and multiple capacitor bank switching, circuit breakers have to interrupt capacitive currents when they are called upon to interrupt the charging currents to overhead lines or cables. Depending upon the design and technology of the circuit breaker, dangerous overvoltages can be produced when capacitive currents are interrupted.

Consider the circuit diagram shown in Figure 6.7. When the circuit breaker in the diagram is closed, the voltage supplied by the generator, V_g, is applied across both the source side and load side capacitors, and the voltages across each V_1 and V_2 are equal in value. However, the current will lead the voltage by about 90° in a purely capacitive circuit. This means that when the current is extinguished by opening the circuit breaker, the voltage V_2 will remain at the value that was impressed upon it by the capacitance, C_2, while the voltage V_1 will follow the generator voltage. A simplified oscillogram showing a successful interruption of capacitive load current is shown in Figure 6.8.

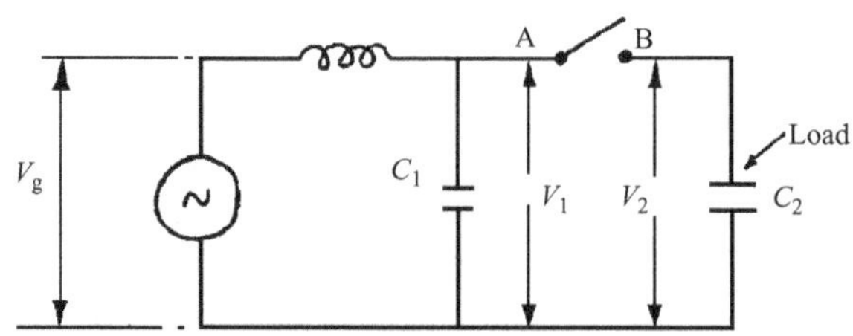

Figure 6.7 Diagram showing a capacitive load circuit

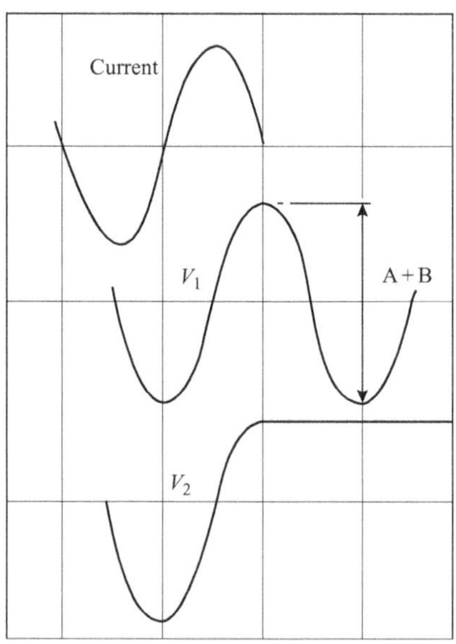

Figure 6.8 Simplified oscillogram of the clearance of capacitive current

Table 6.4 *Typical rated values used for switchgear proving tests*

Rated voltage (kV, r.m.s.)	Line charging current (A, r.m.s.)	Cable charging current (A, r.m.s.)
12	10	25
36	10	50

In the oscillogram shown in Figure 6.8, the top trace is the capacitive current, while the centre and lower traces are the voltages of the circuit breaker terminals A and B. See Figure 6.7. It will be noted that the current shown in the top trace is leading the two voltage traces by 90°. At the time of current interruption, the voltage on the supply-side terminal A will continue to oscillate at the power frequency of the supply. However, within the time scale considered, the load-side terminal B will be maintained at substantially the voltage that was impressed upon the load capacitance at the time of current interruption. As can be seen in Figure 6.8, the voltage across the circuit breaker open contacts, A and B, will then be increased to twice the system peak voltage half a cycle after interruption. As the capacitive current being interrupted is usually very small compared with the short-circuit rating of the circuit breaker, interruption of that current could occur with a very small contact gap. Typical values used for proving the performance of distribution switchgear when interrupting energised overhead lines and cables are shown in Table 6.4. Care should be exercised to ensure that the latest requirements of International and National specifications in terms of current levels and test procedures are used when planning type tests.

A voltage equal to twice the system voltage would stress the small contact gap associated with the interruption of the above small currents and this small contact gap may, or may not, be able to withstand this voltage, resulting in a real danger of restrike taking place. Should this happen, a further interruption would take place, leading to another doubling of voltage. This could eventually lead to an external flashover of insulation and may even result in an evolved fault causing catastrophic failure.

In practice, these excessive theoretical voltages are not obtained. Values between 2.5 and 3.5 times the normal crest voltages are usually the maximum observed and then only on older designs of circuit breaker such as the bulk oil type. The reasons for the limitation seen in practice are as follows:

(1) The capacitor voltage does not remain constant but decays with time, the rate of decay increasing with voltage.
(2) The contact gap would probably break down well before twice the system voltage peak was reached.
(3) The voltage swing is not equal to twice the system voltage but something less, as the capacitor voltage is reduced by the high-frequency oscillation, depending upon the relative values of the supply and the load-side capacitance.

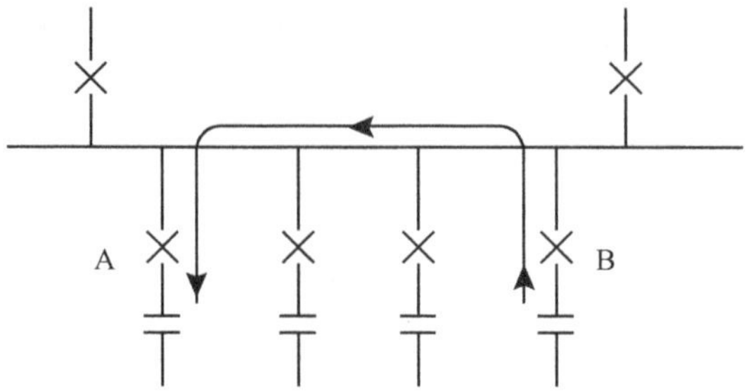

Figure 6.9 Transfer of current between capacitor banks on energisation

(4) The design of the circuit breaker should be such that it is able to recover the dielectric strength of its contact gap as rapidly as possible.

6.6 Back-to-back capacitor switching

When there is more than one capacitor bank connected to a busbar system, the closing of a circuit breaker to energise one bank when the other banks on the busbars are already energised can pose technical difficulties for the circuit breaker due to the high rate of current inrush. The rate of increase in current is far in excess of that experienced in normal terminal fault switching tests. This condition is often met when power factor correction capacitor banks are being switched. Figure 6.9 shows a typical installation having four banks of capacitors. If the circuit breaker controlling bank B is already closed and the circuit breaker controlling bank A is then closed, a rapid transfer of current will flow from capacitor bank B to bank A, as indicated in the diagram. The limitation on the current being the inductance of the busbar system.

Because the rate of rise of charging current into bank A would be in excess of that experienced during normal proving tests, it was the practice to carry out proving tests for the specific application by reproducing the values of capacitance and inductance involved. The possible problem being that the rate of increase of current could be in excess of the rate of increase of contact pressure, leading to excessive contact burning and possible failure. However, the latest version of the applicable International Standard, IEC 62271-100, attempts to establish preferred values of back-to-back capacitor breaking current, inrush current and frequency of the inrush current, the latter determining the time to peak of the inrush current. Selected examples of the preferred values are given in Table 6.5.

The expression for the calculation of capacitor bank inrush current I_{i} and frequency f, when a charged capacitor bank is closed onto an uncharged bank of equal capacitance, is given by:

$$I_{\mathrm{i}} = U_{\mathrm{r}} \times \sqrt{\frac{C}{6 \times L_{\mathrm{b}}}} \qquad (6.2)$$

Table 6.5 Selected examples of preferred capacitor switching values

	Back-to-back capacitor bank switching		
Rated-voltage U_r (kV, r.m.s.)	Rated back-to back capacitor bank breaking current I_{bb} (A, r.m.s)	Rated back-to back capacitor bank inrush making current I_{bi} (kA)	Frequency of the making inrush current f_{bi} (Hz)
7.2	400	20	4250
12	400	20	4250
17.5	400	20	4250
36	400	20	4250

and

$$f = \frac{1}{2\pi\sqrt{L_b \times C}} \tag{6.3}$$

Example 6.4

A circuit breaker controlling the 36 kV power supply to a discharged power factor correction capacitor bank is closed onto energised busbars. These busbars have an identical capacitor bank connected, which is fully charged. Given that the length of the connection is 40 m and the inductance is 1 μH/m, what will be the values of the peak inrush current and its frequency, given that the capacitance of each of the banks is 51.2 μF?

From the data given, it will be seen that the inductance of the connection is $20\,\text{m} \times 1\,\mu\text{H}/m = 20\,\mu\text{H}$. Substituting this value of inductance and the given values of capacitance in Equation (6.2), we have:

$$I_i = 36{,}000 \times \sqrt{\frac{51.2 \times 10^{-6}}{6 \times 20 \times 10^{-6}}}$$

$$I_i = 23.52\,\text{kA}.$$

Similarly, substituting the values in Equation (6.3), we have:

$$f = \frac{1}{2\pi \times \sqrt{51.2 \times 20 \times 10^{-6}}}$$
$$= 4974.5\,\text{Hz}$$

The above calculations show that the magnitude of the inrush current, 23.52 kA, is in excess of the preferred value of 20 kA, while the frequency of the inrush current is well within the preferred value. Under these conditions, it would be beneficial if the inrush current were reduced. The recommended method for carrying this out would be to add inductance between the capacitor banks. By transposing Equation (6.2), and using the value of 20 kA in the expression, the total necessary inductance can be calculated. The added inductance is the calculated figure, less the original natural inductance.

Consideration should be given to the dangers that would be involved if it was possible to close a charged capacitor bank on to energised busbars that had a capacitor already connected. If the point-on-wave of circuit closure was such that the polarities of these capacitor banks were opposite to each other, the resulting inrush current could be doubled. A possible solution in practice would be to ensure that each capcitor bank was fitted with a draining resistor and that a sufficient time delay was introduced for these to be effective between subsequent reclosures.

6.7 Reignition surges

It is technically possible to generate excessively high voltages due to multiple reignitions when switching circuits having significant inductive and capacitive elements with high interrupting ability vacuum interrupters. These high overvoltages are caused by multiple sequential clearing and restriking across the circuit breaker contacts. Vacuum circuit breakers, because of their superior interrupting ability can, under certain circumstances, interrupt high-frequency oscillating currents while their contact gaps are too small to withstand the subsequent impressed recovery voltage that they will experience.

Consider the instance of the contacts of a vacuum circuit breaker separating at, or near, current zero when interrupting the current flowing through a reactor, resulting in a reignition occuring across its short contact gap (see Figure 6.10).

It will be seen that with the recovery voltage shown, the circuit would reignite unless sufficient time was given to allow the gap to increase to that shown at t_r.

When reignition occurs, the initial part of the current that flows will have a high-frequency oscillation due to the natural frequency of the circuit. This current will pass through zero which gives an efficient interrupter the opportunity to extinguish the current. This will trap inductive energy which will be released during the recovery from subsequent interruptions. As the circuit breaker contacts are moving towards the open position, the contact gap increases, and this increases the charge that would be trapped on a subsequent interruption. The cycle of interruption and reignitions will continue, with higher voltages being generated, until the contact gap is large enough for the final clearance to take place. Under certain circumstances, the value of overvoltage produced could be significant. Tests carried out using vacuum interrupters with different contact materials confirmed that the very efficient copper–chrome contacts produced much higher overvoltages than that produced by copper–bismuth. This was because the former was able to interrupt higher frequency currents with a shorter contact gap than the copper–bismuth.

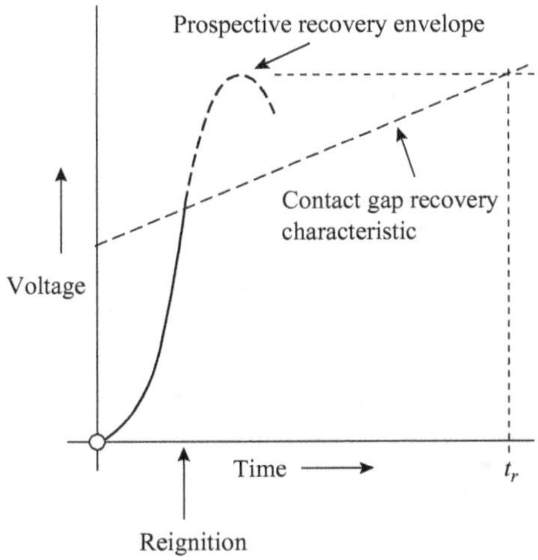

Figure 6.10 Reignition taking place across a short contact gap

Most vacuum interrupter manufacturers produce application guides, which vary in content but may make recommendations with regard to the provision of surge suppressors for capacitor and inductive reactor switching applications. These recommendations should always be followed. If there is any doubt, fit suppressors.

It could be argued that the switching of capacitor banks and inductive reactors is not a normal requirement for distribution switchgear, and that because of this, for normal distribution applications, suppressors are not necessary. However, switchgear engineers should be aware of the possible problems associated with reignition surges associated with these capacitive and inductive switching applications.

Chapter 7
Insulation

It is obvious that conducting components within distribution switchgear have a high voltage potential difference both interphase and with respect to earth when in service. These components have to be securely mounted and fixed in position by using materials which are very poor conductors of electricity. These are known as insulating materials and form a range of components that are continually stressed throughout the whole life of the equipment. It is certain, therefore, that latent defects within insulating materials, due to inadequate selection, design or manufacture, will manifest themselves during the life of the equipment. It follows, therefore, that great care must be taken when designing insulation systems that are to be incorporated within distribution switchgear.

Insulating materials are, by definition, very poor conductors of electricity, which is why they are used to cover conductors, give support to busbars and other conductors, and, in a gaseous form, are used to fill compartments to provide both insulation and a dry, clean environment.

7.1 Electrical stress

An important factor to be taken into account when designing insulation is the electrical stress, both on the surface and through the insulation itself. A good understanding of electrical stress will allow the switchgear designer to ensure that the maximum stress in service never exceeds the permitted level, and also to ensure that the material is used economically. This is achieved by studying the voltage distribution in the area being considered. In the early days, this was carried out by using an electrolytic tank, which was an inflexible and time consuming analytical tool that relied upon the relative resistivity of an electrolytic filled model of the area being considered. Another restriction was that it could only be used for two-dimensional analysis and would not allow the engineer to easily pose 'what if?' questions regarding different shapes and lengths. Nowadays, computer programs exist that will allow both two- and three-dimensional electrical stress analysis to be carried out, with the flexibility to allow shapes to be changed. In addition, these programs will also produce graphs showing the electrical stress on selected planes through the computer model of the area being considered. A typical two-dimensional electrostatic field plot is shown in Figure 7.1.

The example in Figure 7.1 is based on a two-dimensional electrostatic field plot with the centre line for rotation co-incident with the base of the plot. The area studied in the plot is the fixed and moving contacts of an SF_6 puffer circuit breaker. The fixed contact, on the right-hand side, is shown at 100 per cent voltage and the moving contact, on the left, at 0 per cent. The fixed and moving contacts are bridged by a hollow porcelain insulator filled with SF_6 gas and with atmospheric air external to the assembly. A computer program was used to calculate voltage distribution in the arrangement and the plot shows the voltage distribution at 10 per cent intervals. Obviously, the closer these lines are together, the higher the electrical stress. The highest electrical stress is shown to occur at the tip of the fixed contact in the SF_6 gas. A lower, but still reasonably high surface and internal stress occurs on and within the insulating material of the moving contact nozzle. Externally, the stress can be seen to be well distributed, but it should be noted that some of the porcelain sheds towards the left-hand end on the diagram have the same voltage at both their tip and root, and, therefore, do not contribute to the external withstand voltage. These sheds do, however, protect lower sheds from rain and will, therefore, increase the wet withstand voltage of the assembly.

An electrostatic field plot will, therefore, allow the switchgear designer to optimise the thickness and shape of insulation elements by limiting the electrical stress to the value specified by the insulation manufacturer. The value selected should take into account the influence of contamination on insulator performance over the life of the switchgear.

Voltage excursions which occur in distribution networks are caused by switching surges, and depend upon the way in which systems are earthed. For example, an effectively earthed system can be run for 12 h with one phase at earth potential, during which time the equipment will be subjected to a voltage $\sqrt{3}$ times the normal

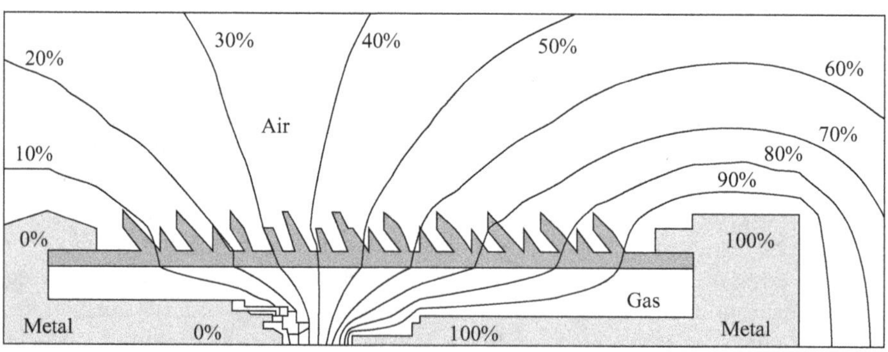

Figure 7.1 An example of an electrostatic field plot [6]

level. The effects of switching surges and system earthing are discussed in Chapter 6. Voltage excursions can lead to electrical discharge within insulation.

7.2 Electrical discharge

Electrical discharge is a precursor to insulation failure and flashover. The electrical discharge mechanism will vary according to the kind of insulation being used. For example, a single electrical discharge in oil, air or SF_6 gas will be followed by full recovery of the insulation; this process is known as 'self-healing'. However, a single breakdown in solid insulation will cause a degree of permanent damage.

7.3 Discharges in oil and gases

When new, insulating oil is a clear, pale yellow liquid. After a period in service, if examination shows the oil to be a darker colour then it is probable that its insulating properties have deteriorated. Impurities in the oil will reduce its insulating properties. In an uniform electrical field, impurities such as fibres and carbon formed by the interrupting process, may form a conducting chain and lead to flashover. However, in a non-uniform field, these impurities will migrate to points of high electrical stress, such as sharp corners. This migration has the effect of rounding-off these sharp points, decreasing the electrical stress and thus increasing the discharge inception and flashover voltage level. Water dissolved in oil is very dangerous, even a 0.01 per cent water content will reduce the electrical strength of the oil by 10 per cent. As oil will float on water, any free water entering a switch tank will accumulate in the bottom of the tank. Not only will some of this free water be absorbed by the oil, reducing its electrical strength, but also in its free state it has been the cause internal flashover and catastrophic failure of switchgear. This is known to have occurred in double-break oil switchgear, and was caused by the moving contact plunger bar drawing the water upwards during the closing stroke. This process is illustrated in Figure 7.2.

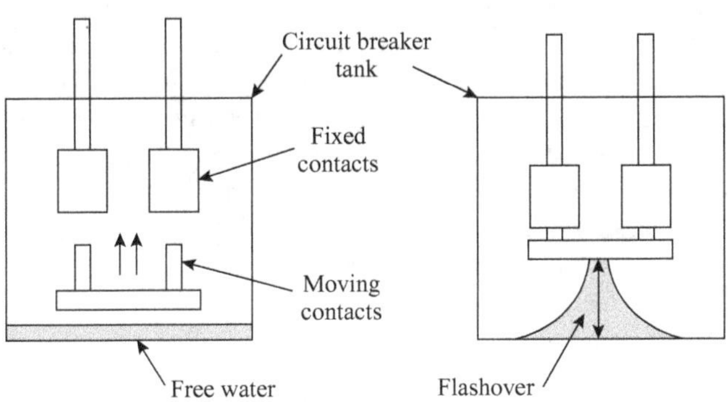

Figure 7.2 Electrical flashover due to free water in an oil circuit breaker tank

When the circuit breaker is open and at rest, as shown in the left-hand diagram of Figure 7.2, the free water in the bottom of the circuit breaker tank will not provoke a flashover. However, during the closing stroke, the contacts will start to move in the direction of the arrows towards the position shown in the right-hand diagram. During this movement, the pressure on the underside of the plunger bar is reduced causing the water to be drawn upwards, bridging the gap between the plunger bar and the circuit breaker tank which is at earth potential, causing an electrical flashover.

If a slowly increasing high voltage is applied across an electrode gap in air, it will be observed that a voltage level will be reached where a glow, known as a corona, will be seen at the sharp edges of the electrodes. This glow will be accompanied by a hissing sound. As the voltage is further increased, the glow will increase in intensity and the sound will change from a hiss to a crackle. As the voltage is increased further, an arc will be struck between the electrodes.

The effect of a flashover is immediate and obvious, but partial discharge taking place at lower voltages can also lead to flashover, so early investigation and corrective action must be taken. Apart from measurement with ultrasound or electrostatic equipment, there are three obvious signs of lower voltage level electrical discharge taking place. These are the smell of the ozone produced, the sound of hissing or crackling and corrosion of the ferrous components and fastenings. This corrosion is a result of the formation of nitrous oxide and nitric acid, which are produced by the discharge activity. For example, when entering a multi-panel substation and noticing the smell of ozone and corrosion taking place on one panel, it should be relatively easy to identify that this panel is experiencing an electrical discharge problem.

7.4 Discharge in solid insulation

The situation with regard to solid insulation is very different. A single breakdown will permanently damage the insulation, lowering its insulating value. Further breakdowns will continue to increase the damage, eventually leading to failure at normal system voltage. Solid insulation should, therefore, always be designed with a much higher margin of safety. It follows, therefore, that insulation design should be such as to minimise electrical discharge. A successful design requires that knowledge of the voltage levels for discharge inception and discharge extinction is determined. (Partial electrical discharge testing is discussed in Chapter 13.) When the applied voltage is increased, a level is reached where electrical discharge can be detected which is known as the discharge inception level. If the voltage is subsequently reduced, the discharge will be extinguished, and this is known as the voltage extinction level. This will always be lower than the discharge inception level. A design philosophy needs to be established which sets limits for these two values.

7.5 Discharge level design practice

It is essential that discharge extinction is higher than the maximum system voltage, otherwise there will be a risk of eventual insulation failure. A good practice would

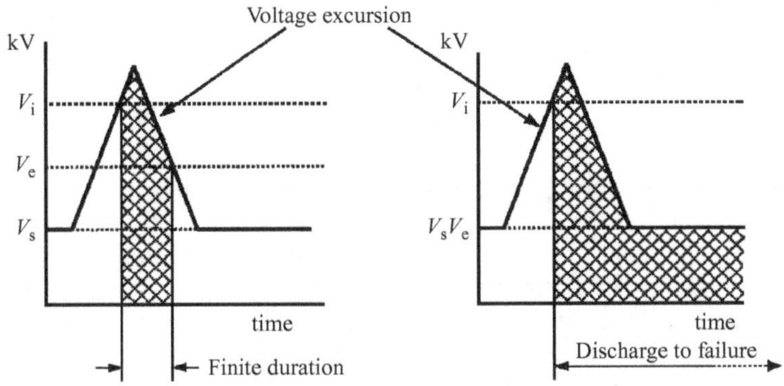

Figure 7.3 Diagram of electrical discharge voltage levels

be for there to be a margin of at least 10 per cent between the maximum system voltage and the discharge extinction level. Discharge inception and extinction levels are illustrated in Figure 7.3.

In the diagrams shown in Figure 7.3, the voltage is plotted vertically and time horizontally. The symbols shown are: (i) V_i, discharge inception voltage, (ii) V_e, discharge extinction voltage and (iii) V_s, system voltage.

The diagrams illustrate how the relative value of extinction voltage influences the duration of electrical discharge following a voltage excursion. In the left-hand diagram, electrical discharge is shown to be initiated as the value of the excursion voltage reaches the inception voltage. The voltage excursion, which could have been the result of a switching surge, reaches a peak and then returns to the normal system voltage. However, before it reaches the system voltage, it falls below the discharge extinction voltage level and this will result in the cessation of electrical discharge activity. The electrical discharge is, therefore, of finite and short duration. The right-hand diagram illustrates what happens if the extinction voltage is equal to, or less than the system voltage. Under these conditions, electrical discharge will continue until insulation failure takes place.

7.6 Voids in moulded insulation

Unless precautions are taken, voids may be formed within cast resin insulators during the casting process. These voids will be detected during the routine high-voltage discharge tests that should be carried out on each insulator. The problem is in deciding what is acceptable and what is not. Unless the casting process was carried out in a vacuum or SF_6 gas, the voids will be filled with air. There are schools of thought that believe a discharge within a void will cause the air/resin boundary of the void to become conducting, eventually shorting out the void and causing the discharge to cease. However, any electrical discharge within an air-filled void will produce nitric acid and could lead to eventual insulation failure. A prudent insulation policy would be to allow a maximum discharge level of 5μ coulombs at 1.5 times the phase voltage with an extinction level of at least 1.1 times the phase voltage. 5μ coulombs was selected as this was considered to be the minimum reliable discharge reading.

7.7 Flashover caused by indirect discharge

Before leaving the subject of electrical discharge and eventual flashover, it is important to understand that insulation failure and flashover can occur even if the insulator is correctly designed and tested as a component. Actual examples of two such cases of insulation failure are described in the following.

The first concerned an outdoor design of vacuum circuit breaker in India. A significant number of these units were installed in service, and a worrying number of them failed only six weeks later. The three-phase vacuum unit concerned consisted of an air-filled steel enclosure fitted with porcelain bushing in the roof, for connection to an overhead line. A vacuum interrupter module was located within the steel housing, supported by fibreglass insulators. At the outset, it was realised that all of the units would be located in a heavy rainfall area, so no covers or doors were allowed except on the underside of the unit.

A diagrammatic cross section of the unit is shown in Figure 7.4. The flashovers experienced were down the module support insulators. It was suspected that electrical discharge was taking place within the unit so discharge measurements were made on a number of units in a laboratory capable of very sensitive measurements. No problems were found with discharge inception and extinction levels. It was realised that the tests carried out had not included simulation of the very high humidity that was present at the time of the failures, so the bottom covers were opened and steam from a kettle was allowed into the unit which was then energised at system voltage. Almost immediately, the air gap between the roof bushings and the steel enclosure lit up like neon lights with an electrical discharge. The source of the problem, therefore, was identified as discharge within the small air gap between the roof bushings and the enclosure and the products of this discharge caused the failures of the support insulators. Verification of the cause of the problem was obtained by adjusting the

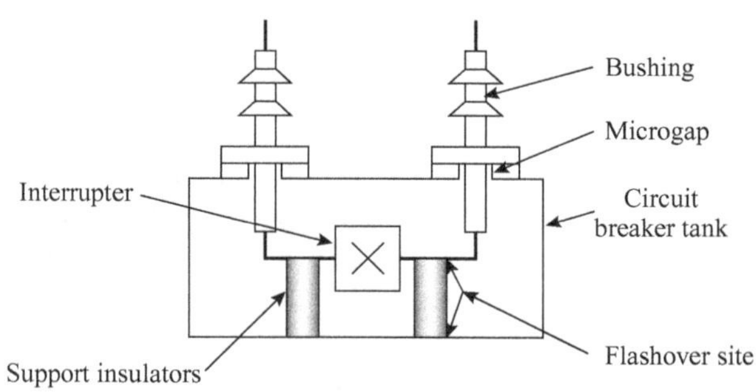

Figure 7.4 Indirect cause of flashover

air gap. Following this modification on all units, no further problems were experienced. Two lessons were learned from this experience. First, small air gaps, known as microgaps, between insulation on conductors and earthed material should always be avoided, and second, the insulation performance should be verified by simulating the anticipated environmental conditions.

A second case of indirect discharge causing flashover involved vertically isolated switchgear. Large numbers of this type of switchgear were installed in two adjacent electrical utilities in the Far East. A number of units in one of the utilities experienced insulation failures of wound-type current transformer supports, while units in the other utility were trouble free. As all the failures were in only one of the utilities and both utilities had the same climate conditions, environmental causes could be ruled out. In addition, as it was known that production of circuit breakers for both utilities was mixed, it was fair to assume that the problem was not one of manufacturing quality. It was considered that the problem was associated with the utility's installation of the circuit breakers. Details of the installation procedures of both utilities showed that the trouble-free utility only used bituminous compound-filled cable boxes, whereas the utility with the problem only used heat shrink cable terminations. The heat shrink terminations were the suspected source of the electrical discharge that resulted in the support insulator failures. Subsequent electrical discharge site measurements confirmed that a high percentage of heat shrink terminations had sustained discharge at system voltage. This confirmed that the support insulator failures were the indirect result of discharge elsewhere in the unit. Figure 7.5 shows a diagrammatic cross section through the unit.

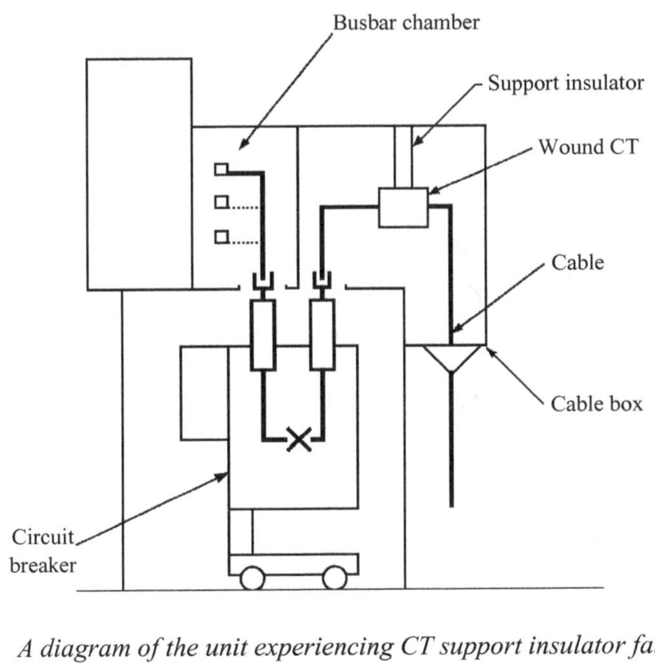

Figure 7.5 A diagram of the unit experiencing CT support insulator failures

The utility's investigation into the cause of the flashovers found that, due to inadequate training of the cable fitters, they were cutting the semi-conducting layer of the terminations too short. As a result of this experience, the utility instituted two actions to prevent a recurrence. The first action was the immediate introduction of a revised training programme for all cable fitters, and the second was to introduce discharge measurements on all units, present and future, on a regular basis using hand-held ultrasound detectors.

7.8 Breakdown voltage and gas pressure

There is a relationship between gas pressure and breakdown voltage which is given by Paschen's law. This was discussed in Chapter 2 in relation to the high levels of vacuum pressure as used within vacuum interrupters, at which this relationship is not linear. However, within the pressure range of gas and air used for insulation purposes in distribution switchgear, the gas pressure and maximum withstand voltage level have a linear relationship which, for a uniform electrical field, is defined by the formula:

$$V_{\text{withstand voltage}} = P_{\text{gas pressure}} \times D_{\text{electrode gap}}.$$

This relationship for air and SF_6 is shown in Figure 7.6 and compares these with transformer oil. All are based upon a 12.5 mm sphere to plane electrode system with

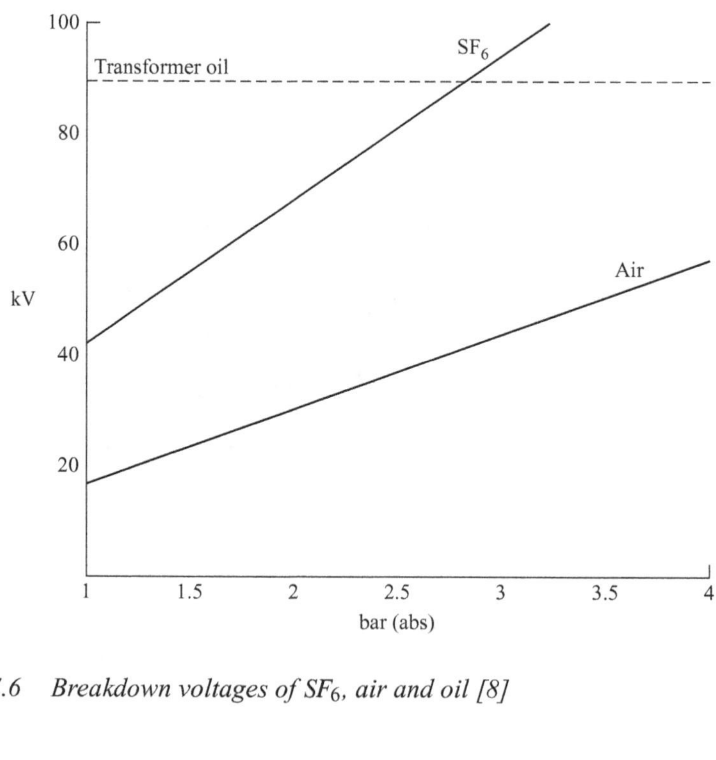

Figure 7.6 Breakdown voltages of SF_6, air and oil [8]

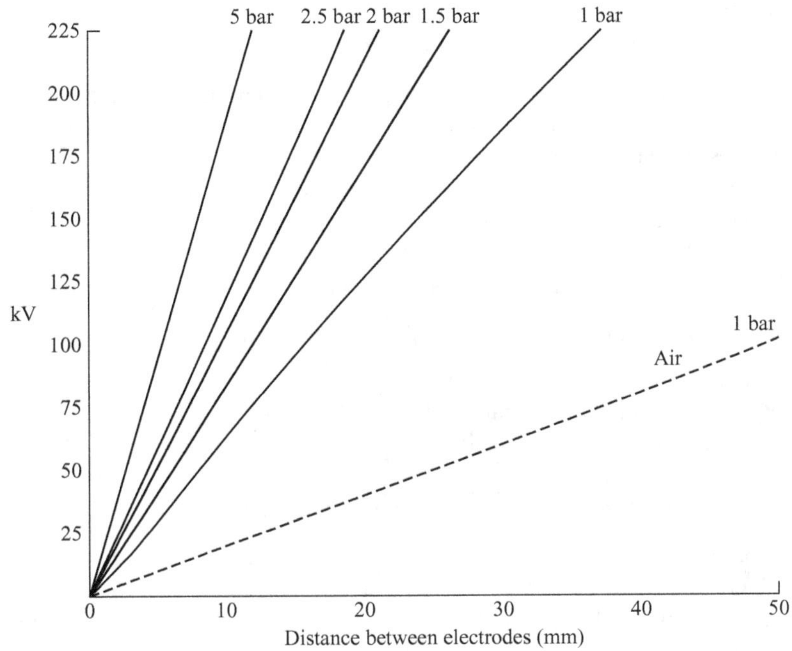

Figure 7.7 Influence of electrode gap upon breakdown voltage [9]

a 12.5 mm gap. Similarly, the influence of electrode gap on the breakdown voltage in a uniform field is demonstrated in Figure 7.7.

7.9 Solid insulation

Solid insulation will permit the passage of a very small current when an alternating voltage is applied. The small current that flows will produce losses in the insulation material, which will reduce its electrical resistance. The reduction in resistance will cause the small leakage current to increase, producing further losses and a further reduction in resistance. This, in turn, will give yet a further increase in leakage current, and the process will continue until either the temperature for chemical change and failure of the insulation material is reached, or equilibrium is achieved.

It should be noted that breakdown voltage is not directly proportional to the insulation thickness. This is probably due to the thicker material not being able to dissipate internally produced heat as well as thinner material. The breakdown voltage of solid insulation is given by Baur's law which states that

$$V = dt^{2/3}$$

where V is the breakdown voltage, d, a constant for the material and t, the thickness of the material.

Apart from a high dielectric strength, the desirable characteristics of solid insulation can be said to be the following:

(1) Long term reliability. At least 40+ years of withstanding electrical stress.
(2) Ability to withstand the maximum working temperature combined with electrical stress without change.
(3) Mechanically strong, as insulators are commonly subjected to mechanical stress from both mechanical operations and the passage of peak short-circuit currents.
(4) Freedom from harmful effects on the environment.
(5) Freedom from 'end of life' disposal problems.

Historically, high-voltage insulation has progressed from:

Porcelain, slate, mica and rubber,
Porcelain and bakalised paper (BP),
Porcelain, BP and epoxy resin,
Epoxy resin and BP,
↓ Epoxy resin, glass filled epoxy, polyurethane resin (PU).

Solid insulation in modern switchgear is usually based on epoxy or polyurethane resin. Where electrical properties have also to be combined with high mechanical strength, glass fibre or kevlar reinforcement is used. Typical examples of this type of reinforcement can be found in drive links from the operating mechanism to the interrupter, and in pressure enclosures.

Until recent times, epoxy resin filled with silica flour was used for moulded components. The filler helping to reduce shrinkage during cooling as well as increasing the electrical anti-tracking properties of the insulation.

Cast epoxy and polyurethane resin insulation has to be carefully designed to minimise locked-up internal mechanical stress that may be produced during the cooling and contraction process when casting has taken place. This is done by avoiding sudden changes in cross-section and using a generous radius wherever a section change takes place. Any insert, for example, those used for fastening purposes, should be relatively large in diameter and be hemispherical at their buried end. Conductors can pose a potential problem, as contraction of the resin will take place towards the centre of its mass and conductors will generally run the length of the moulded resin. The stress generated by shrinkage on a conductor will be minimised by controlled knurling of the conductor (see Figure 7.8).

The two methods used to minimise locked-up mechanical stress induced by an embedded conductor are shown in Figure 7.8. The first method is to knurl only one end of the conductor so that it is gripped by the resin, freeing the remainder, coated in silicone grease, to allow it to slide as the resin cools. The second method is to knurl the whole length of the embedded conductor so that it is gripped by the resin along its whole length.

If these measures are not properly carried out, then thermal shock, or any additional mechanical stress could amplify the locked-up stress and cause a fracture in the insulating material with disastrous consequences (see Chapter 13, Section 13.3.4.3, short time current tests).

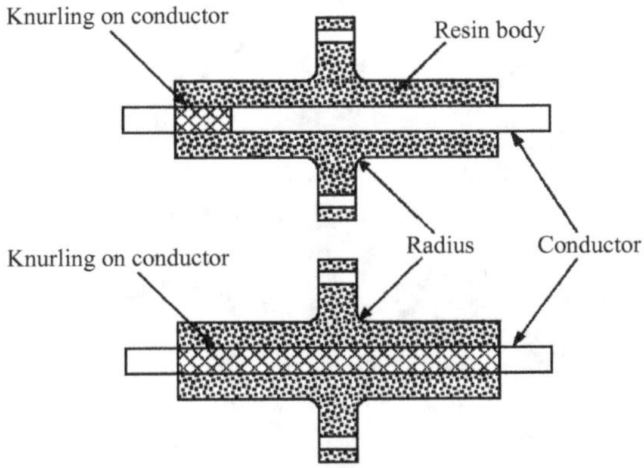

Figure 7.8 Design of a resin insulated bushing

Figure 7.9 Complex moulding (courtesy of Medelec Switchgear Ltd Malta)

In addition to the routine tests specified in standards, some manufacturers carry out a useful and searching test on randomly selected mouldings. This test subjects the mouldings to a low-temperature soak at about −30°C followed by a visual inspection to look for small fractures, before carrying out the normal routine tests. This low-temperature soak confirms that the preparations and manufacturing procedures are correct.

The production process for epoxy mouldings is lengthy in that the tool cycle time can be up to 24 h. If production quantities are significant, then a large number of tools will be required, making the whole process very expensive. This can be contrasted with the modern polyurethane mouldings which, with pressure gelation,

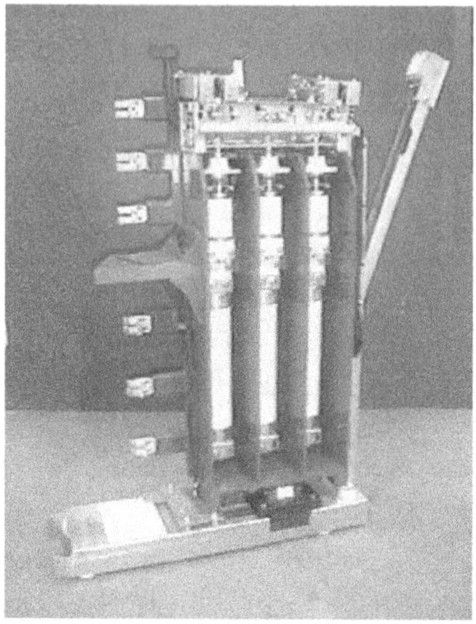

Figure 7.10 Large moulding (courtesy of ALSTOM T&D Ltd)

can be produced in only about 30 min. Complex and large mouldings in polyurethane are now produced in large quantities. Examples can be seen in Figures 7.9 and 7.10.

Even though polyurethane moulding material is more expensive than epoxy cast resin, the short cycle time of polyurethane means that the cost per moulding produced in this material is by far the most economical and this is now the preferred manufacturing process for quantity production.

7.10 Composite insulation

There are situations in the design of distribution switchgear where there is more than one type of insulation used in combination between a live conductor and earthed metal. These composite types of insulation could be as simple as a conductor embedded in a resin moulded bushing, with air insulation between the stem of the bushing and the earthed metal compartment walls. In cases like these, an electrostatic field plot as shown in Figure 7.1 would be used to evaluate the electrical stresses within both the resin and the air.

However, there are instances where relatively long conductors and insulators are used which are clad in two types of solid insulating material. These types of applications can be analysed by simple calculation.

Consider the case of several layers of insulation, each layer having a different dielectric constant. Each layer can be considered to behave like a capacitor, therefore:

$$\frac{1}{C} = \frac{1}{C_1} + \frac{1}{C_2} \ldots \text{ etc.}$$

The charge on each capacitor will be the same:

$$q = C_t V_t = C_1 V_1 = C_2 V_2 \ldots \text{ etc.}$$

and

$$V = \frac{q}{C}.$$

Therefore

$$\frac{V_1}{V} = \frac{C}{C_1} \cdots \tag{7.1}$$

$$\frac{V_2}{V} = \frac{C}{C_2}$$

and so on.

The importance of this is that it demonstrates that calculations can be made to determine the electrical stress in each layer of multi-layered insulation. The electrical stress gradient will not be uniform across the insulation thickness as it will be different in each layer depending upon the dielectric constant for that material.

Consider the case of a 12 kV conductor that is insulated from earth by 10 mm of solid insulation. If a single homogeneous insulating layer is used, the electrical stress will be independent of the dielectric constant of the insulation and will have a linear voltage gradient. This voltage gradient will be equal to

$$\text{Gradient} = \frac{12{,}000\text{ V}}{10\text{ mm}} = 1200\text{ V/mm}.$$

If, however, three layers of insulation were used, having, say, dielectric constants of 6, 2 and 4, with thicknesses of 3, 3 and 4 mm respectively, the gradient for each layer will be different as they will be similar to series connected capacitors.

The capacitance of each layer can be calculated by using the expression for parallel plate capacitors:

$$C = \frac{kA}{4 \times \pi \times d}$$

where k is the dielectric constant, A, the area of the plates (cm^2) and d, the distance between the plates (cm).

Therefore given that in layer 1 $d = 3$ mm and $k = 6$, the capacitance will be:

$$C_1 = \frac{6 \times A}{4 \times \pi \times 0.3} = 1.59\text{ A}.$$

Similarly

$$C_2 = \frac{2 \times A}{4 \times \pi \times 0.3} = 0.53\text{ A}.$$

and

$$C_3 = \frac{4 \times A}{4 \times \pi \times 0.4} = 0.796\ \text{A}.$$

The total capacitance of all three layers is then calculated from the expression:

$$\frac{1}{C_t} = \frac{1}{C_1} + \frac{1}{C_2} + \frac{1}{C_3}$$

and multiplying both sides by $1/A$ becomes

$$\frac{1}{C_t A} = \frac{1}{C_1 A} + \frac{1}{C_2 A} + \frac{1}{C_3 A}$$

Therefore

$$\frac{1}{C_t} = \frac{1}{1.59} + \frac{1}{0.53} + \frac{1}{0.796} = 3.772$$

and

$$C_t = \frac{1}{3.772} = 0.265.$$

From Equation (7.1), the voltage across the first insulation layer will be given by:

$$\begin{aligned} V_1 &= \frac{C_t \times V_t}{C_1} = \frac{0.265 \times 12{,}000}{1.59} \\ &= 2000\ \text{V}. \end{aligned}$$

The voltage stress, f, in the first layer will be

$$\begin{aligned} f_1 &= \frac{V_t}{d_1} = \frac{2000\ \text{V}}{3\ \text{mm}} \\ &= 666.7\ \text{V/mm}. \end{aligned}$$

The voltages and the electrical stress within the other two layers can be calculated using the same method, giving the following values:

$$V_2 = 6000\ \text{V} \quad \text{and} \quad f_2 = 2000\ \text{V/mm}.$$

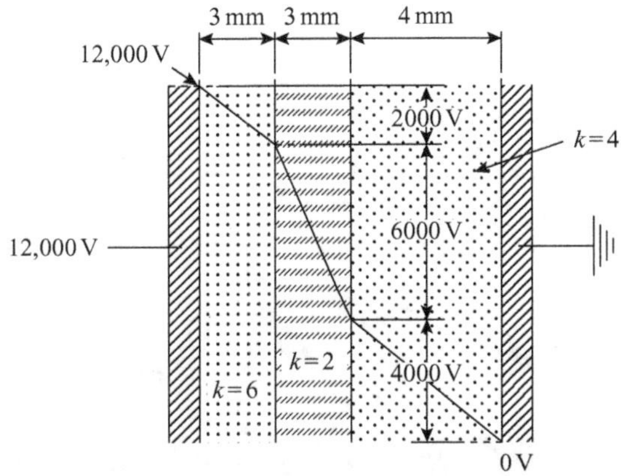

Figure 7.11 Physical arrangement and voltage gradients in composite insulation

Similarly,

$$V_3 = 4000\text{ V} \quad \text{and} \quad f_3 = 1000\text{ V/mm}.$$

The physical arrangement and voltage gradients are shown graphically in Figure 7.11.

The calculated voltage gradients are shown superimposed upon the dimensions of the composite insulation shown in Figure 7.11. It will be seen that the voltage gradient across the insulation layer with the lowest dielectric constant is greater than that across materials having higher dielectric constants.

Chapter 8

Operating mechanisms

It is the function of a circuit breaker operating mechanism to transmit stored energy via a mechanical drive to the moving contacts, so as to cause them to close and open, when commanded, within defined operating times and speeds. What is more, it should operate without hesitation even after prolonged periods of inactivity. Operating mechanisms will incorporate drives to ancillary devices such as auxiliary switches for remote control and indication, motor drives for spring charging, position indicators and local manual trip and close facilities. In many cases, an operations counter will also be required.

8.1 Materials

Special attention has to be paid to the material specification and design of shaft and latch bearings in order to ensure operation without hesitation even after very prolonged periods of inactivity. Under these circumstances, the moving contacts should still part within about 25 ms when the protection system issues an opening command. Trip and close latches are intentionally designed to be relatively lightly loaded in order to minimise trip and closing coil burdens. To achieve this, they have a high mechanical advantage. However, this brings with it a high velocity ratio with resultant high-impact speeds of the latch faces. It is, therefore, common practice to use case hardened steel for the latch faces to prevent deformation which would lead to changes in tripping load and, therefore, operating time.

The intended service environment of the switchgear will have a marked influence on the choice of materials and protective finishes. All components within operating mechanisms, such as levers, links and side sheets, need to be able to withstand more than 20 years of service without corrosion within their intended service environment. This can be achieved by either using inherently immune materials, such as stainless steel or aluminium bronze, or by electroplating. However, electroplating of springs should be avoided as it can cause premature failure due to hydrogen embrittlement. Intergranular corrosion should also be guarded against by applying a proprietary spring protection coating as soon as possible after manufacture. Advice on suitable coatings can be obtained from specialist spring manufacturers.

8.2 Operating features

The following operating features will be found in many specifications:

(a) *Trip free*
The operating mechanism must be able to trip at any part of the closing stroke, and the trip signal will always take precedence over a closing signal. The exception to this is the mechanical drive to a puffer type interrupter which, for arc interruption purposes, must complete its closing stroke before opening in order to recharge its interrupting cylinder with gas.

(b) *Independent operation*
Sufficient power must be applied independently of the operator to fully close or open under all conditions.

(c) *Electrical trip*
For local and remote operation, including protection tripping.

(d) *Manual trip*
Local operation only.

(e) *Electrical close*
Usually only for remote operation.

(f) *Manual close*
Local operation only.

8.3 Energy for operation

The energy to operate the mechanism can be stored in a number of different ways, including:

(1) *Springs:* A very common energy system still widely used.
(2) *Electrical:* Also a common system, which is now growing in numbers with the increasing use of magnetic actuators.
(3) *Potential energy:* Derived from a suspended weight (see Figure 11.3).
(4) *Compressed gas:* Has been used for higher voltage distribution circuit breakers at 66 kV and above, and also for the operation of pole mounted disconnectors. In this latter application, the compressed gas was in the form of commercially available nitrogen cylinders.
(5) *Chemical:* Is still used for trip initiation within HRC fuse striker pins (see Chapter 12 and Figure 12.1). However, in this form, it does not provide the actual energy for operation. Some research work was carried out into the use of aero engine chemical starter cartridges to provide operating energy, but these did not prove to be commercially viable.

In this chapter, only the spring and magnetic actuator types of circuit breaker operating mechanisms will be described, as they are the most common types in use today.

8.4 Spring operating mechanisms

The use of a relatively low powered solenoid trip coil to release high levels of stored energy is normal for spring-operated mechanisms. This is carried out by the use of a trip latch. There are two types of trip latch that will be discussed; these are the 'D' trip latch and the roller toggle latch. Figure 8.1 shows a 'D' trip latch.

In Figure 8.1, the spring load, L, is acting on a lever at a radius of R_1 tending to move the lever in an anti-clockwise direction. This movement is restricted by the face of the lever being in contact with the 'D' latch at a mean radius of R_2 and the face of the 'D' latch acting at a radius of R_4. The trip solenoid, when energised, provides a tripping load T acting at a radius of R_3 to cause the 'D' latch to rotate in an anti-clockwise direction to release the spring force L when the 'D' latch clears the face of the latch acting at radius R_2.

If we ignore the frictional resistance of the bearings of the 'D' latch and lever, we can easily calculate the minimum tripping load that the trip solenoid would experience.

$$\text{The force between the latch faces} = \frac{L \times R_1}{R_2}.$$

$$\text{Given that the co-efficient of friction} = \mu$$

$$\text{Minimum tripping load} = \frac{\mu \times L \times R_1 \times R_4}{R_2 \times R_3}.$$

The actual co-efficient of friction will probably be of the order of 0.1–0.3.

Because of the inherently high friction between the latch faces, the use of the 'D' type trip latch will generally be confined to operating mechanisms employing a relatively low level of spring force.

For high-loaded operating mechanisms the tripping loads can be minimised by employing a toggle trip latch, an example of which is shown in Figure 8.2.

Unlike the 'D' trip latch described, the latch faces of the toggle trip latch shown in Figure 8.2 do not have a sliding interface. Therefore, the frictional load will be significantly lower. The toggle trip latch is much better suited to being used in operating mechanisms having a relatively high spring load. The lower link, having a length of A has the mechanism spring load L acting on its lower axis pin. As this link has an axis pin at each end, it can only transmit load in line with its two axis pins. As the

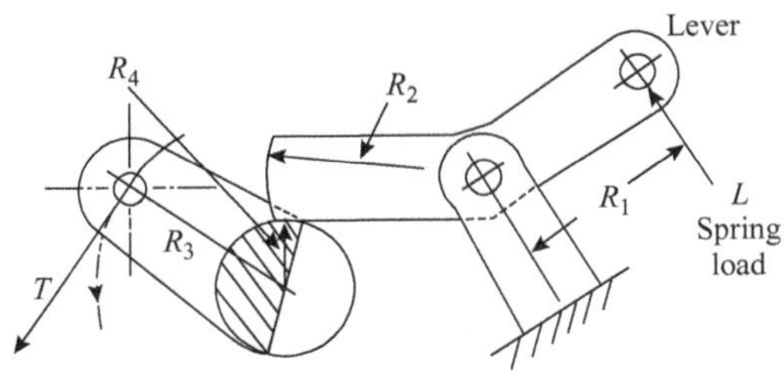

Figure 8.1 A 'D' type trip latch

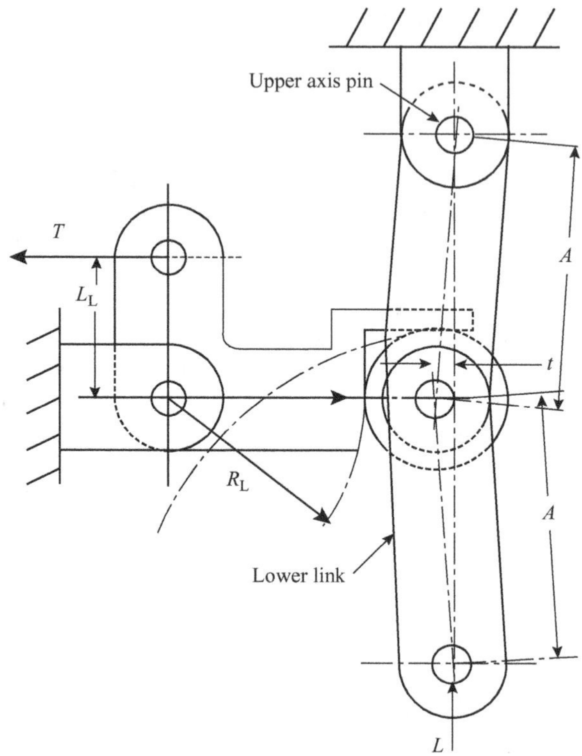

Figure 8.2 Toggle trip latch

upper axis pin is offset by the distance t, the latch roller will be maintained in contact with the latch face by a component of spring force (see Figure 8.3).

The dimensions and the offset of the lower link are shown in Figure 8.3 on the left, and the resulting forces and their directions are shown on the right-hand side. The force acting on the latch face t_L can be determined by applying the rules of similar triangles:

$$\frac{t_L}{L} = \frac{t}{A}.$$

Therefore

$$t_L = \frac{L \times t}{A}.$$

Given that the co-efficient of the latch rolling friction is μ_r, the minimum tripping load required from the trip solenoid, T, can be determined from the dimensions of the latch levers and links shown in Figure 8.2.

$$T \times L_L = R_L \times \mu_r \times t_L$$

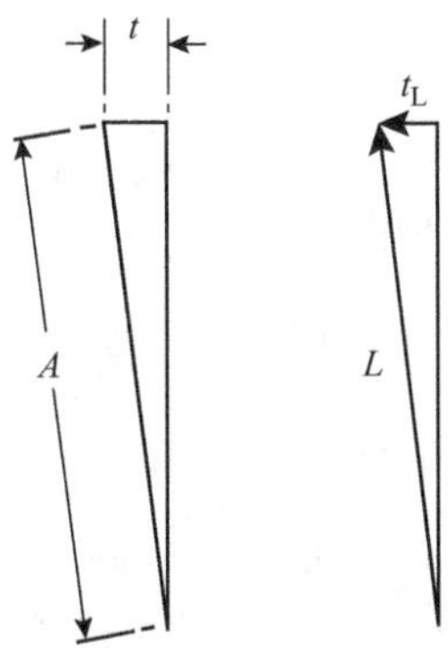

Figure 8.3 *Dimensions and vectors of forces acting on the latch face*

Therefore

$$T \times L_L = \frac{R_L \times \mu_r \times L \times t}{A}$$

and

$$T = \frac{R_L \times \mu_r \times L \times t}{A}.$$

The tripping load would, of course, be increased by the frictional resistance of the pin bearings and this should be taken into account.

As the latch roller is in contact with a latch face that has a constant radius of R_L, it will maintain a constant latch load. The trip solenoid, not shown in Figure 8.2, acts by applying a tripping load of T at a radius of L_L. Latch stability is provided by a light spring, not shown, acting in the opposite direction to the tripping load T. In practice, the latch roller would normally be arranged to have a case hardened active face and, to minimise friction, it would also run on a needle roller bearing. In addition, some designers ensure that the left-hand latch component is dynamically balanced by arranging for the mass of latch metal to be equally spread on either side of its bearing so that a mechanical shock in any direction will not cause the latch to rotate. This is particularly important when switchgear is to be located in an area that is subjected to seismic shock.

The hinge points shown as being fixed in Figure 8.2 may be arranged to be fixed only when the operating mechanism is in a certain condition, for example, when it is reset, or when it is reset and its springs are charged.

A word of warning before leaving the subject of latches. As a general rule, the armature of trip and electrical release solenoids should never be lubricated. Always consult the manufacturer's product handbook. Numerous instances of failure to trip or failure to close on electrical release have been caused by over-enthusiastic lubrication in the solenoid armature region, even though specific warnings against doing this were given by the manufacturer in the product handbook. The reason for the warnings is that, with time, the lubricating oil will dry out leaving a sticky residue which prevents the solenoid armature operating.

8.5 Three-link kinematic chains

The closing force profile requirements of circuit breaker operating mechanisms is such that a relatively low force will be required during the initial part of the contact stroke in order to give a smooth start to the acceleration of the moving parts. The operating mechanism must provide a relatively high force towards the end of the closing stroke in order to overcome frictional drag of contact loading imposed by the electrical contact system. These requirements are not compatible with the load characteristics provided by helical closing springs, which start very high and fall away as the spring expands. A three-link kinematic chain is used to provide the operating mechanism force required from a helical closing spring by selecting crank angles and dimensions in order to change the output load during the operating cycle. An example of this is shown in Figure 8.4.

The three-link kinematic chain shown in Figure 8.4 consists of a load input crank A a connecting link B and an output crank C. The relationship between both output torque and velocity ratio for the three-link kinematic chain illustrated is detailed in Figure 8.5.

The output crank is arranged to receive the drive thrust from link B at a relatively small radius. This will convert the high start spring load to a small start torque. It is also arranged to act at a large radius at the end of the stroke, where the spring load will be at a minimum. The input crank is shown to have a start angle, α, of 50° and a final angle of 10°. This final angle leaves the driven end of link B with an offset of t which is an almost in-line position, imparting a high output torque.

Latch and three-link kinematic chain elements described earlier which are used in typical spring-operated mechanisms can be found in Figure 8.6, which is a line

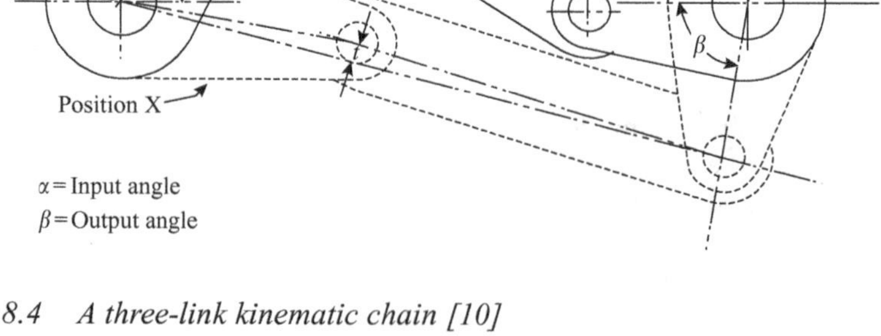

Figure 8.4 A three-link kinematic chain [10]

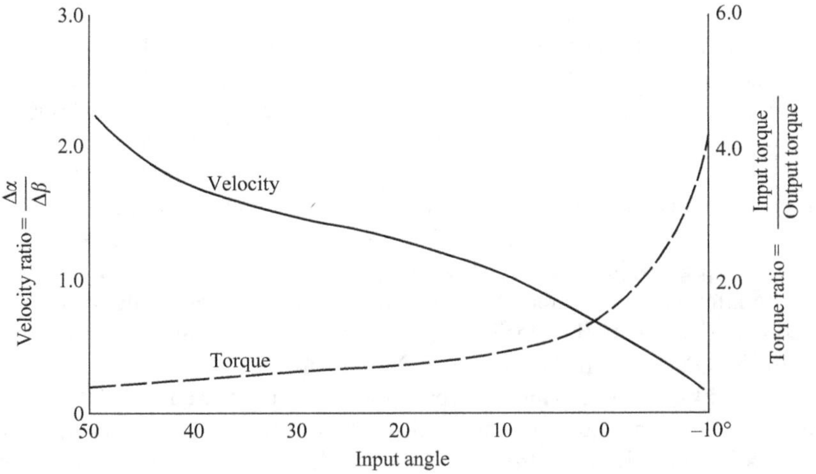

Figure 8.5 Torque and velocity relationship of three-link kinematic chain [10]

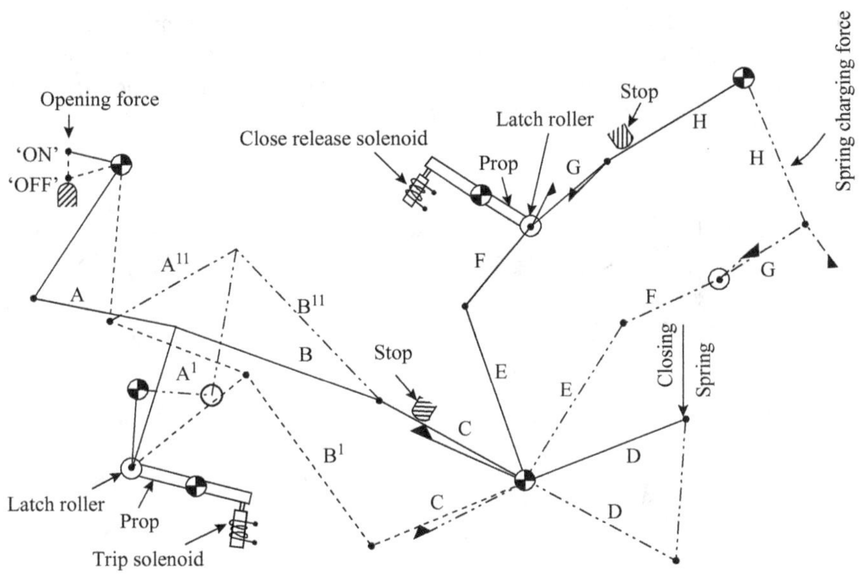

Figure 8.6 Line diagram of a spring mechanism having an electrical release [10]

diagram of the linkages within a mechanism having a trip coil for opening and an electrical release solenoid for closing. It will be appreciated that the design of spring-operated mechanisms is an involved and complex operation. It will usually be found that the design operation is carried out by experienced specialists within a switchgear manufacturing company, who will use historical data, established design elements, materials, surface finishes and hardness criteria that have been proven by service experience and time. In addition, extensive use of high-speed photography will be made to examine all aspects of the mechanism's performance in detail.

The dotted and chain dotted lines in Figure 8.6 indicate the link positions that are taken up when a trip or close operation has been carried out. The solid lines show the positions taken when the circuit breaker is closed and its springs have been charged. Examination of Figure 8.6 shows that clockwise rotation of crank H to its fixed stop position will cause the closing spring to be charged, and the Close release solenoid latch to be engaged. It will be noted that links F and G are arranged to be not quite in line. This toggle alignment imparts a thrust on the release coil latch roller so that when the latch prop is rotated clockwise by the release solenoid the links F and G will collapse, releasing the closing spring energy and causing an operation to take place to the ON position. When in the ON position, a similar toggle alignment is formed by links A and B which causes a load to be impressed by the trip solenoid roller on to its prop face. This means that when in this condition, energisation of the trip solenoid will cause the links to collapse to the A^1 and B^1 positions and the circuit breaker to move to the OPEN position. Also shown are link positions A^{11} and B^{11} which are the positions that the trip linkage takes up if a trip signal is received when a closing operation is in progress, thus meeting the requirement for a trip operation having preference over a close operation. Examination of all of the linkages shows examples of three-link kinematic chains to harness the spring loads to the circuit breaker operation requirements. The physical embodiment of the mechanism described is shown in Figure 8.7.

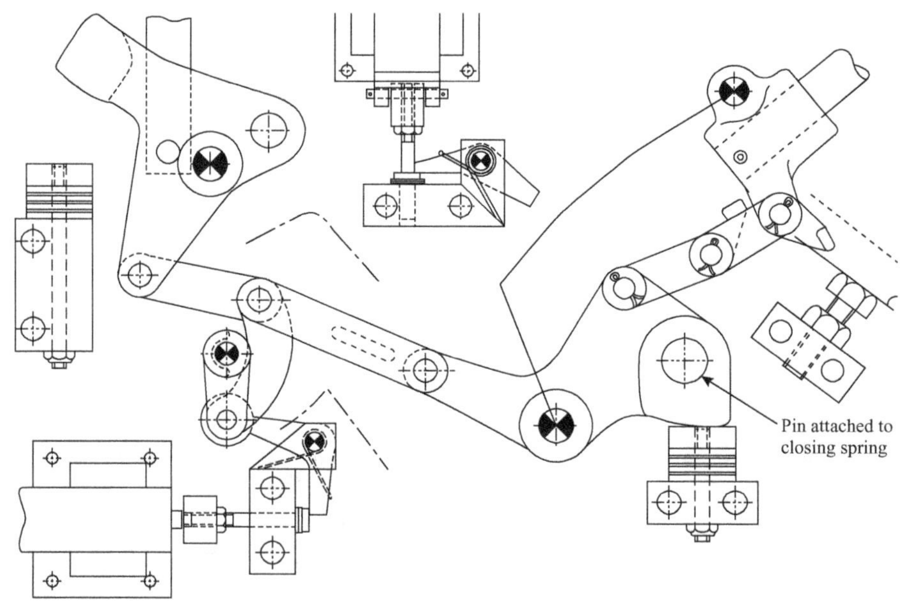

Figure 8.7 Physical arrangement of linkages in the mechanism described above [10]

8.6 Magnetic actuators

Common sense, confirmed by numerous surveys including an extensive international survey by CIGRE in the 1980s, indicates that the more components there are that go to make up an assembly, the greater the potential for unreliability. This is particularly true when considering components that move, such as those found in operating mechanisms. It follows, therefore, that if a new concept for an operating mechanism was conceived which offered virtually only one moving part, it would be taken up in various forms by a high proportion of switchgear manufacturers. This was the case with the magnetic actuator. The principle of operation of the magnetic actuator can best be described by reference to Figure 8.8 which shows the typical construction found in most magnetic actuators.

It can be seen that a typical magnetic actuator has a central armature housed within a magnetic yoke. This armature, when compelled, will move from the lower position shown in the diagram to its upper position, and then vice versa. The travel being defined by the space between the ends of the armature and the end walls of the magnetic yoke. Around a section the armature, and in close proximity, is a set of permanent magnets which are arranged to have poles laterally arranged across their shortest length. As shown in the diagram, the flux from these magnets links the armature, the air gap and the magnetic yoke. While some magnetic actuators operate by using a single coil, Figure 8.8 is shown to be a two-coil arrangement. These coils are wound so that, when energised, they will produce magnetic flux as shown in Figure 8.8 when starting in the OPEN position. It can be seen that this flux opposes the flux produced by the permanent magnets at the bottom end of the actuator, releasing the magnetic hold that was in place, and augments the flux produced by the permanent magnets across the air gap at the top of the actuator. If sufficient flux is produced,

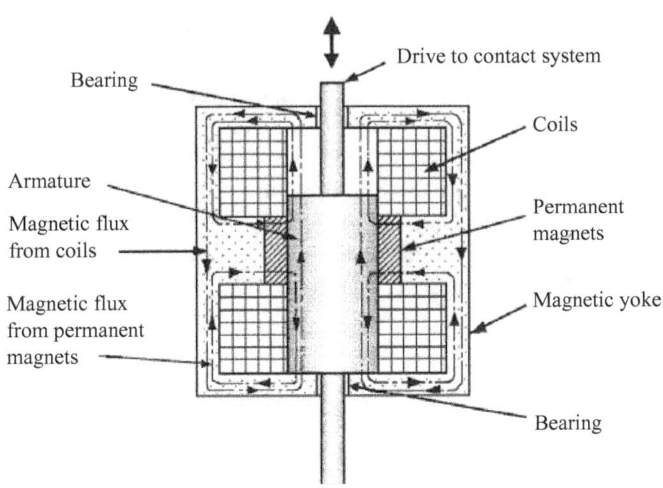

Figure 8.8 Typical construction of a magnetic actuator

Figure 8.9 Cylindrical form of magnetic actuator (courtesy of ALSTOM T&D Ltd)

the armature will move to the upper or ON position and will be magnetically held in this position when the operating coils are de-energised. When in the ON position, current passing through the coils in the reverse direction will, by similar action to that described for closing, cause the actuator to move to the OPEN position. Production magnetic actuators can be either cylindrical in form, or a flat rectangular box-like form as shown in Figures 8.9 and 8.10.

In the 1960s, a low VA trip coil was patented by George F. Chrisp [11], which was mechanically charged and held in the charged position by permanent magnets. In many respects, this device can be said to have anticipated today's magnetic actuator, but the energy it was able to release was limited by the strength of the magnets which were available at that time. However, considerable progress has been made in magnet technology since that time and this has led to the widespread application of magnetic actuators with vacuum interrupters in distribution switchgear. The vacuum interrupter being essentially a short stroke device as is magnetic actuator making them an ideal combination. The progress in magnet technology can be seen in Figure 8.11.

One of the attractions of the magnetic actuator is that it consumes no power when the circuit breaker is in the ON or OFF position. However, it must hold the contacts in either position with sufficient force to meet the operational requirements. While the actuator is essentially a bi-stable 'flip/flop' device it must do more than that, in that it must not only hold the contacts in the OFF position during normal operations, but it must also maintain them in that position when subjected to external shock loads which form part of its service site conditions. It must also hold its vacuum interrupters in the ON position with a force at least equal to the minimum external force specified by the vacuum interrupter manufacturer which is designed to allow the contact system to safely carry its maximum rated peak and short time short-circuit current. In other words it must maintain a designed minimum hold-on load.

Typical minimum hold-on forces specified by vacuum interrupter manufacturers are given in Figure 8.12. However, when designing a magnetic actuator, information relating to the specific vacuum interrupter that is intended to be used must be obtained.

Figure 8.10 *Rectangular form of magnetic actuator (courtesy of Medelec Switchgear Ltd)*

The magnetic actuator must provide a hold-on force greater than the minimum specified for three vacuum interrupters. This is because vacuum interrupters have butt contacts which must have a maintained contact load in the closed position even when the maximum contact erosion has taken place.

The principle of the interrupter drive arrangement is shown in Figure 8.13. This shows that the mechanism drive to the interrupter is by a linking element which provides the final drive via a spring. This spring provides automatic compensation for both contact wear and differences in contact alignment. As the lowest applied spring load should be that recommended by the vacuum interrupter manufacturer, it follows that the actual applied load will be at a maximum when the interrupters are new, with no contact erosion, and must be at least equal to the minimum recommended load plus the spring rate multiplied by the erosion allowance.

As it is usual to drive all three interrupters with one magnetic actuator, the minimum hold-on force of the actuator will be equal to this new higher load.

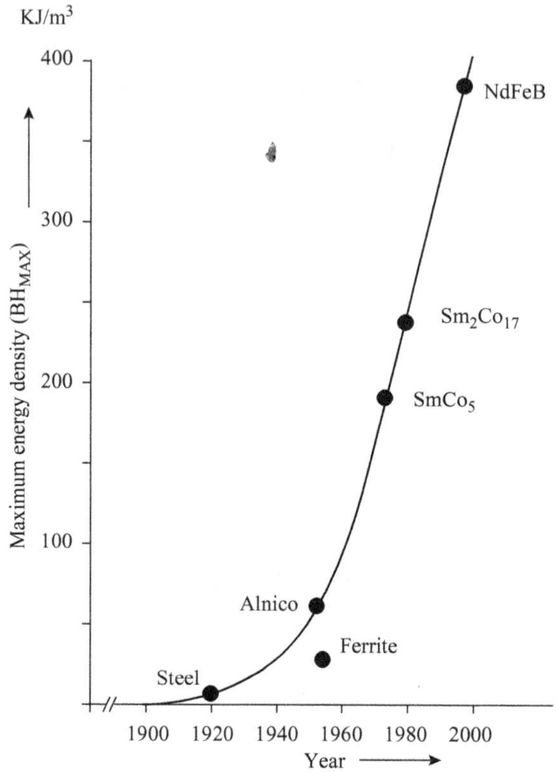

Figure 8.11 Development of magnetic energy density in the twentieth century [12]

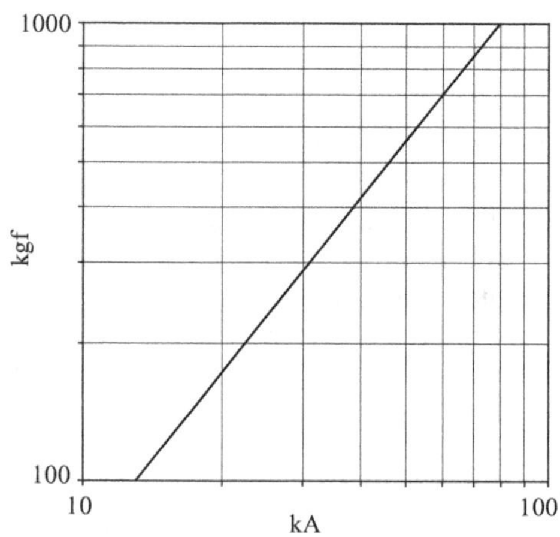

Figure 8.12 Typical relationship between vacuum interrupter contact force and kA rating [13]

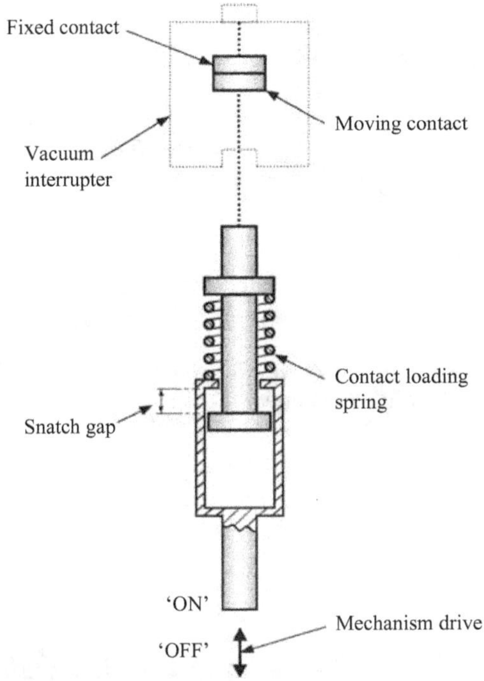

Figure 8.13 Principle of vacuum interrupter drive

The stages to be followed in designing a magnetic actuator are:

(1) Determine the minimum hold-on force.
(2) Decide if the actuator drive is to be direct. (It is assumed for this exercise that it is.)
(3) If (2) is direct, the stroke of the actuator will be equal to the interrupter stroke plus the snatch gap allowance for contact erosion. Typically, a vacuum interrupter will have contact gaps and snatch gaps of 8 and 5 mm for 11 kV and 12 and 6 mm for 36 kV applications.
(4) Determine the dimensions of the permanent magnets to provide the hold-on force. The force produced by the permanent magnets alone can be predicted and measured for given movements of the armature. A typical profile of these forces is shown in Figure 8.14, which relates to an actuator having an armature travel of more than 20 mm.
(5) Decide upon the voltage and internal impedance of the power source that will energise the actuator.
(6) Decide upon the dimensions of the magnetic yoke.
(7) Calculate the Ampere-turns of the operating coils that will overcome the hold-on force.

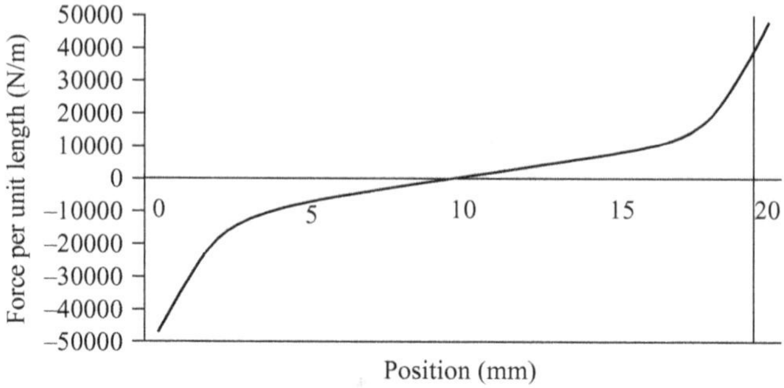

Figure 8.14 *Static armature force-stroke characteristic (courtesy of ALSTOM Medium Voltage Switchgear, South Africa)*

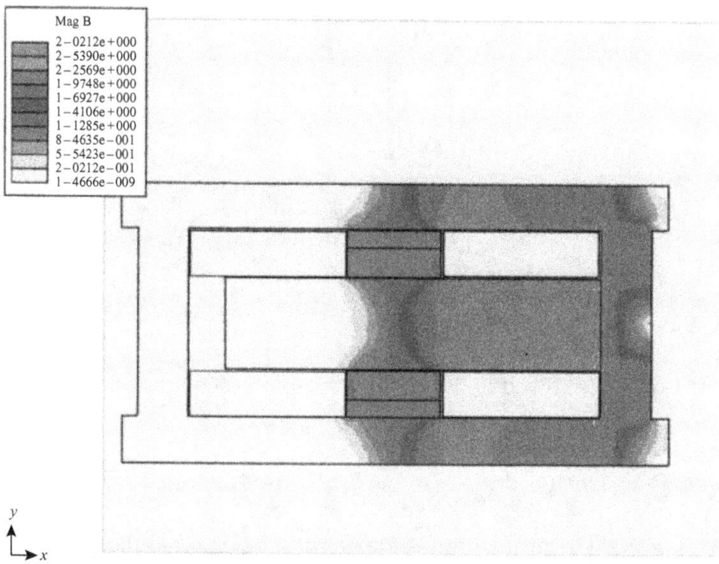

Figure 8.15 *Dynamic analysis of magnetic flux (courtesy of ALSTOM Medium Voltage Switchgear, South Africa)*

This process may well be iterative and will need to be repeated for various degrees of actuator travel. However, there are specialist companies that can assist by producing a four-dimensional analysis (three physical dimensions plus time) that will indicate the performance of the actuator modelled in real time. This will allow the design to be understood and refined to exactly match the operational requirements of the application. This, in turn, will have an important influence on the efficiency of the actuator design and also upon its cost, as the cost of the actuator is directly related to the energy it can supply [5] (see Figure 8.15).

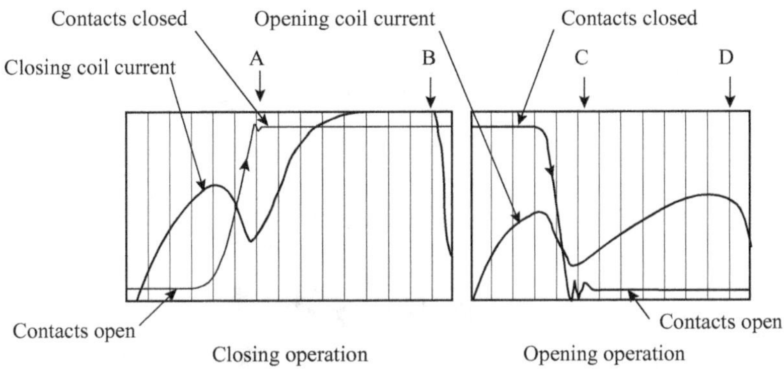

Figure 8.16 Closing and opening operation oscillograms

Having an efficient actuator of minimum cost will still leave a potential cost issue with regard to its power source when applications such as stand-alone secondary switchgear equipment are concerned. This is because the power source is typically Lithium batteries which can have a significant cost. Consider typical closing and opening records of such a device, which are shown in Figure 8.16.

It will be seen in the oscillograms shown in Figure 8.16 that both the closing coil and opening coil currents show an initial rise in value and then a fall. This is due to the change in inductance that takes place as the armature of the actuator changes its position. Once the contacts complete their travel, shown as points A and C, the coil currents continue to rise until they attain their maximum value. Auxiliary switches driven by the main contacts will then interrupt the coil currents at points B and D. There are two disadvantages associated with using this method of coil current interruption.

(1) As the actuator is magnetically latched at each end of its travel, the passage of further current serves no operational purpose. The energy drawn from the batteries after contact movement ceases is therefore wasted and the battery used must be much larger than would otherwise be necessary.
(2) If, for some reason, the current passing through the coils is reduced to a sub-operation level, coil burn-out could occur.

The use of a timing device in the coil energisation circuit would avoid the second disadvantage but would still involve the use of a larger battery than would otherwise be necessary.

A means of detecting the second rise in coil current would minimise the size of battery needed, or alternatively, as patented by Johnson and Dilkes [14], proximity switches incorporated within the actuator (and energised at the same time as the coils) would also allow much smaller batteries to be used.

Chapter 9
Primary switchgear

As the name implies, primary switchgear is the first stage in the process of conducting electrical power from the grid to the end user. The importance of the strategic position of a primary substation and its switchgear within the system means that the layout, design and operation must ensure maximum availability and reliability.

As the system impedance is lower at the primary substation than further into the network, the fault level tends to be higher, usually between 25 and 50 kA.

Until the 1960s, primary switchgear was invariably of the oil type, with most installations being very large by today's standards. An example of the size and complexity of these substations can be seen in Figure 9.1, which shows a Ferguson Pailin type VRP, 33 kV, 13.1 kA substation installed in the Swansea area in 1930.

The scale of the switchboard in the photograph can be judged by the relative size of the figure in the foreground. This oil switchgear, which was equipped with duplicate busbars, can be compared to its modern vacuum equivalent, shown in Figure 9.2. The modern equipment is much less than half the size and has more than twice the short circuit rating of its predecessor.

Although the manufacture of oil switchgear in the UK has been discontinued in favour of non-oil types, there are still large numbers in use and, over the years, they have given good service.

While duplicate busbar schemes are still used by some industrial users and overseas power generation companies, they are rarely specified for today's primary switchgear. This is because insulation reliability has greatly improved with modern moulding materials, and the cost of single busbar protection schemes has been reduced by the use of the blocking features of microprocessor-based relays.

The substation arrangement of primary switchgear must be given careful consideration in order to provide continuity of supply in the event of a transformer or incomer switchgear fault. This is usually achieved by enhancing the application flexibility through a measure of equipment redundancy. Considerable flexibility of supply in primary substations with single busbars is usually obtained by having the primary transformers connected to busbars which can be coupled using a bus-section circuit breaker, as the typical arrangement shown in Figure 9.3.

Figure 9.1 A Ferguson Pailin type VRP circuit breaker installed in 1930 (courtesy of ALSTOM T&D Ltd)

Figure 9.2 Type WSB 33 kV duplicate busbar vacuum switchgear (courtesy of ALSTOM T&D Ltd)

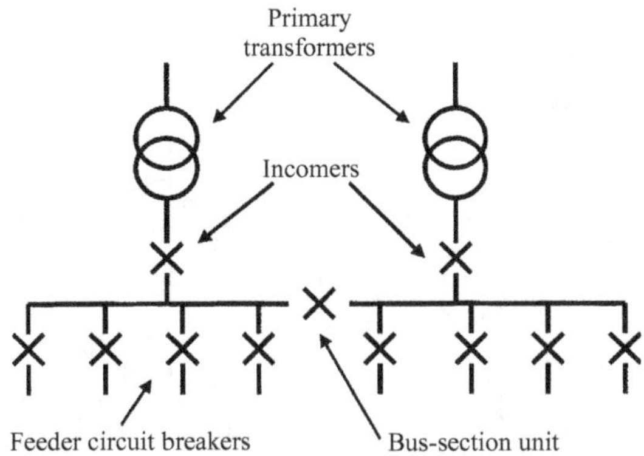

Figure 9.3 Primary substation with two transformers and a bus-section circuit breaker

In the event of a fault in one of the primary transformers shown in Figure 9.3, supply could be restored to all of the outgoing feeder circuit breakers by opening the incoming circuit breaker and closing the bus-section unit. Withdrawable and interchangeable circuit breakers would enhance availability in an emergency by enabling rapid replacement of a faulty circuit breaker.

9.1 Changes in technology

The changes that have taken place in all types of switchgear are the result of two principal drivers. The first being active competition, and the second changes in specifications. The former brings about change by encouraging manufacturers to invest in R&D to achieve reductions in the cost and size of their switchgear, as well as increasing the ratings and facilities that they can offer to customers. The latter is a mandatory change which manufacturers must adopt.

Let us examine the R&D effort first. A large percentage of switchgear development costs is spent on equipment certification and a major authority for carrying out independent short-circuit tests is the KEMA laboratories in the Netherlands. KEMA is independent of all manufacturers and has established an international reputation for the quality of the proving tests it carries out in order to award a certificate of rating. As a result of this, many of the world's leading manufacturers use KEMA as a certifying body. Assuming that the work carried out by the KEMA laboratories represents the overall relative activity of manufacturers, then an analysis of the certificates the laboratories have issued from 1985 to 1997, for switchgear in the range of 12–36 kV, will indicate the relative effort being made by manufacturers on different interrupting technologies. Such an analysis is shown in Figure 9.4.

Figure 9.4 shows the number of certificates issued by the KEMA laboratories [5] plotted vertically, for vacuum and SF_6 types of interrupter, against the years plotted horizontally. It will be seen that the ratio of vacuum to SF_6 certificates is about 10:1.

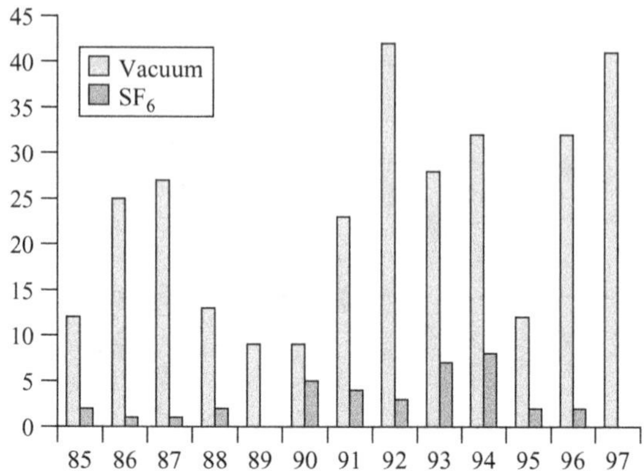

Figure 9.4 Analysis of 12 kV to 36 kV KEMA circuit breaker certificates [18]

Figure 9.5 The evolution of 20 kA vacuum interrupters through the 1960s, 1970s, 1980s and 1990s (courtesy of ALSTOM T&D) [18]

Clearly manufacturers have been putting considerably more effort into the development of vacuum interrupter-type switchgear than into SF_6. The reason for this is that manufacturers have been, and still are, making considerable progress with the design and manufacturing technology of vacuum interrupters and the associated circuit breaker components. Having said that, it is important when making this comparison to distinguish between switchgear which uses SF_6 for insulating purposes and switchgear which also uses the gas for interrupting purposes. The former, known as GIS form, is gaining a significant share of the available market. A striking example of the results of continuous development can be seen in the advances that have been made in vacuum interrupter technology. Figure 9.5 shows the progressive reduction in the size of vacuum interrupters for a given rating that has taken place over the last four decades.

The dramatic reduction in size that can be seen is the direct result of investment in R&D, and mirrors the overall reduction in switchgear size seen in Figures 9.1 and 9.2.

9.2 Current and voltage transformers

Important peripherals, such as current transformers (CTs) and voltage transformers (VTs) have also seen radical design changes as a result of the work done to introduce epoxy and polyurethane moulding materials.

Ring-type current transformers have been used for many years in the UK-manufactured equipment and are now being introduced into switchgear manufactured on the continent. The reason for this can be found in the relative cost ratio of the two, which is between three and five in favour of ring-type current transformers. Ring-type current transformers can be readily clamped to resin-insulated bushings within the current transformer chamber, as the surface of the moulded insulation on the primary conductor can be given an earth layer by using silver loaded paint or hot metallic zinc spray. Similarly, this system can be used to prevent electrical discharge between electrically stressed insulation and its surrounding metal enclosure. Figure 9.6 shows a modern voltage transformer cast in polyurethane with a metalised surface.

Figure 9.6 Moulded interior of a 12 kV voltage transformer (courtesy of ALSTOM T&D Ltd)

The three vertical projections that can be seen in Figure 9.6 are housings for high-voltage VT fuse-links. The value of using high-voltage fuse-links to protect the voltage transformer must be questioned. The lowest high-voltage fuse rating that can be produced is 3.15 A. It is known that, under ferro-resonance conditions, a current of between 400 and 600 mA will flow and cause the voltage transformer to fail; the fuse, therefore, provides no protection under these conditions. The argument for fitting VT fuses is that anything fitted to high-voltage equipment must be protected. The counter-argument is that protection is already provided by the circuit breaker and that the fuse itself, and its accommodation, may actually cause problems in service. It is known that continental switchgear manufacturers do not normally fit VT fuses and the latest published edition of the Electricity Association Standard, EATS41-36, allows the user to choose whether or not to fit VT fuses.

Alternatives to conventional current and voltage transformers, using (1) Ragowski, (2) Hall effect and (3) capacitive devices hold out the promise of desirable features for the switchgear designer, allowing greater freedom in the architecture that can be chosen, but to date, these devices have not made appreciable inroads into the market. The reason for this may be due to user conservatism or a lack of enthusiasm because of the need for additional electronic signal amplifiers.

9.3 The architecture of primary switchgear

There are many possible ways in which the essential components of primary switchgear, that is, the busbars, circuit breaker, cable boxes and means of earthing, together with protection and instrumentation, can be arranged. Switchgear designers have used their ingenuity to arrive at arrangements which offer the user certain features that are considered to be advantageous in terms of increasing the availability of their primary switchgear. The following examples illustrate some of these features.

9.3.1 Horizontal transfer earthing

The way in which the essential components are arranged within switchgear employing vertical isolation and earthing, with horizontal transfer between the functions is shown in Figure 9.7.

The circuit breaker element is mounted on wheels and includes a vertical racking mechanism. The diagram shows the circuit breaker in the raised position, linking the busbars and cable. The dotted lines show the circuit breaker in the busbar earth and the cable earth positions. It should be noted that when in either of these two positions, automatic trip protection is disabled. Interlocking must be provided to prevent the circuit breaker closing or opening unless in a fully engaged or fully isolated position.

A photograph of typical vertically isolated horizontally withdrawn switchgear is shown in Figure 9.8. This form of switchgear was prevalent in the UK and UK-influenced markets for many years. Originally incorporating oil circuit breakers, some

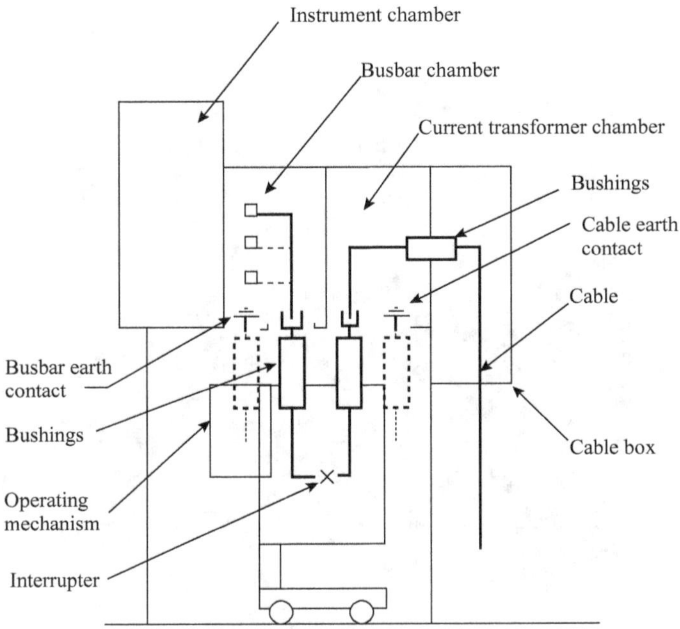

Figure 9.7 Arrangement of components within vertically isolated switchgear

types, such as that shown above, survived by conversion to vacuum interrupters in air. The advantages of this type of switchgear include:

(a) The need for a separate earthing device and associated interlocks is obviated.
(b) The position of the circuit breaker, in either the isolated or the engaged position, is obvious. Similarly the position with regards to the busbar earth, the service position, or the cable earth position can be readily seen when entering the substation.

9.3.2 Horizontal isolation with separate earthing switches

Separating the functions of fault current interruption and earthing, by the use of separate devices, was common practice within Continental Europe and is currently widely employed in the UK. A typical diagram of the internal architecture of this type of switchgear is shown in Figure 9.9.

It will be seen in the diagram shown in Figure 9.9 that the circuit breaker is a wheeled structure with primary isolating contacts arranged to engage with contacts within a fixed housing. A separate cable earthing switch is located within the fixed housing. Not shown in the diagram are the earthed shutters that would normally cover the fixed primary isolating contacts when the circuit breaker is moved out of engagement. Also not shown is the interlock system, which would prevent the earthing switch being operated with the circuit breaker in engagement. Busbar earthing

Figure 9.8 Type VMX vertically isolated horizontally withdrawn switchgear (courtesy of ALSTOM T&D Ltd)

normally requires another earthing switch to be fitted, but some users will accept the deliberate removal of this interlock, to allow the busbars to be earthed through the circuit breaker when the earth switch is in the closed position. Under these conditions, automatic protection tripping of the circuit breaker would be disabled.

As with all isolatable switchgear, interlocks are also provided to prevent operation unless the circuit breaker is fully engaged or isolated. The isolation distance is specified as the distance which will withstand electrical breakdown when impressed with 115 per cent of the equipment's rated impulse voltage. Standards also require that the withdrawable portion of the equipment remains connected to earth potential until this the isolating distance is reached. A photograph of a circuit breaker of this type is shown in Figure 9.10.

9.3.3 Horizontal isolation with internal earthing via vertical transfer

Switchgear employing horizontal isolation with vertical transfer for internal earthing via the circuit breaker is very similar to the horizontal transfer switchgear described earlier and shown in Figures 9.9 and 9.10.

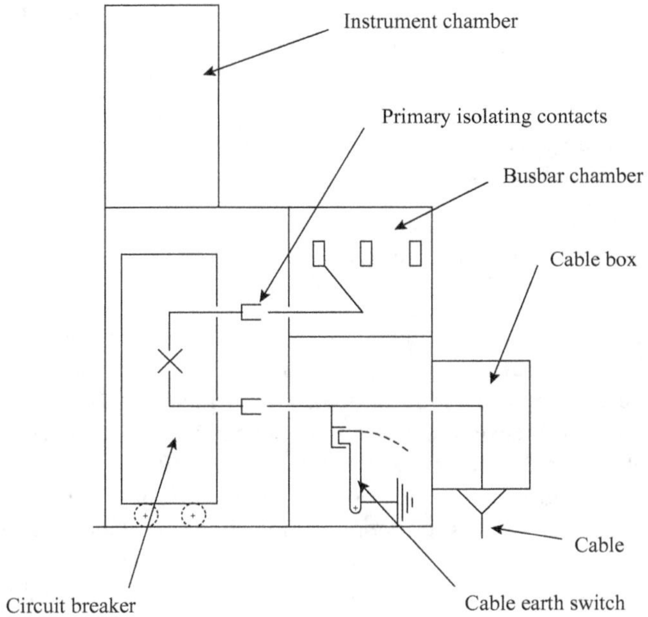

Figure 9.9 Horizontally isolated circuit breaker with separate earth switch

Figure 9.10 Horizontally isolated circuit breakers having separate earth switches; type HWX (courtesy of ALSTOM T&D Ltd)

In this version, the circuit breaker element is carried on a truck which incorporates a screw jack for raising and lowering the circuit breaker. When inserted in the fully raised position, the circuit breaker can be used to earth the busbars; in the central position, the circuit breaker links the cable to the busbars and in the lower position, it can be used to earth the cable. The three vertical transfer positions are shown in Figure 9.11.

The advantages of this arrangement are:

(1) No earth switches are required for earthing the busbars or cable.
(2) The position of the circuit breaker in either the busbar earth, service or cable earth positions is obvious to the operator.

The disadvantage of the arrangement, from the manufacturers' point of view rather than that of the user, is that the expense of equipping every circuit breaker with racking equipment to earth the busbars has to be met. This facility is not required on more than, say, two circuit breakers on any one switchboard. A switchboard of this type, nearing completion in the factory, is shown in Figure 9.12.

It can be seen that there are two ports at the bottom of the front sheet of each circuit breaker. One of these is for horizontal insertion and withdrawal of the circuit breaker, and the other is for vertical positioning in either the busbar earth, service or circuit earth positions. The method of pre-selection and indication of these positions within the fixed housing is shown in Figure 9.13. It will be noted that, for operational security, facilities are provided to padlock the selector in any one of the three positions.

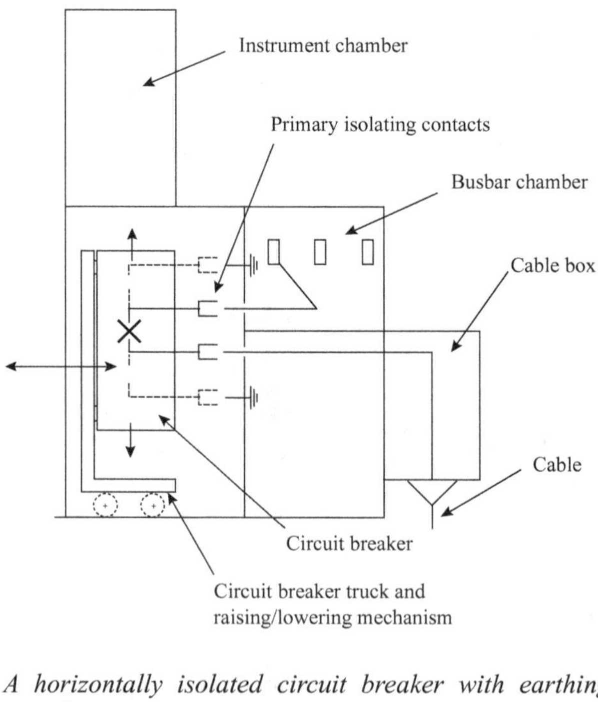

Figure 9.11 A horizontally isolated circuit breaker with earthing via vertical transfer

Figure 9.12 *Type MV12 switchgear, horizontally isolated with earthing via vertical transfer (courtesy of Medelec Switchgear Ltd Malta)*

Figure 9.13 *Pre-selector of circuit breaker position; type MV12 switchgear (courtesy of Medelec Switchgear Ltd Malta)*

9.3.4 Horizontal isolation with internal earthing via top contact stem rotation

The final form of primary switchgear earthing to be described is one that is unique and has been used successfully for many years. This one again uses the circuit breaker to apply the earth, but does so by rotation of the circuit breaker top connector stems so that they engage with earth contacts within the fixed housing. A diagram of this arrangement is shown in Figure 9.14. A photograph of this type of switchgear is shown in Figure 9.15.

The ingenuity of switchgear designers is always being challenged to bring advantages that will make their product more attractive to the user. It was recognised that the overall size of switchgear together with the space required for installation is of importance to the user, as the total cost is a function of substation size. Switchgear designers studied three areas where it was considered that space could be saved. The first was the space occupied by a circuit breaker when it was isolated from its primary connections, particularly if this was to be carried out behind a closed cubicle door. The second was the space necessary for withdrawal of the circuit breaker from the cubicle and the third was the space at the rear of the switchgear to give access for connecting the power cables. One such design was produced which made large contributions to savings in all three areas. This was the VISAX, produced by ALSTOM T&D Ltd which is shown in Figures 9.16 and 9.17.

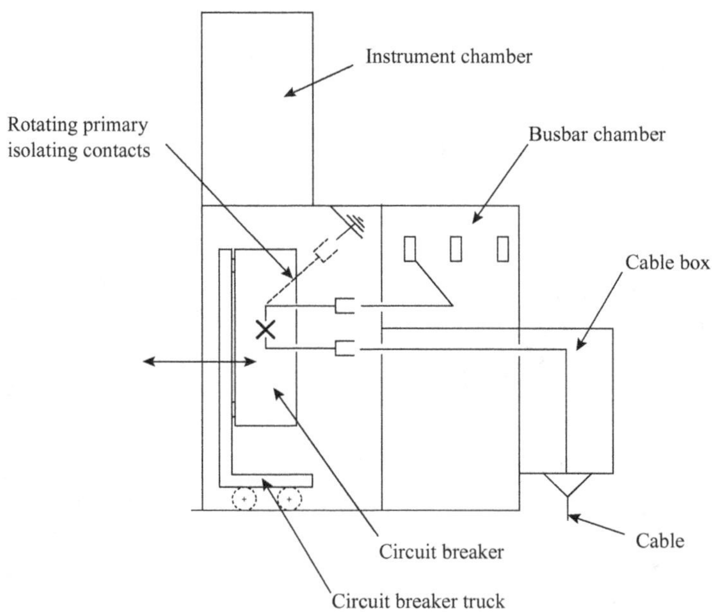

Figure 9.14 Elements of a circuit breaker having horizontal isolation with cable earthing via top isolating contact stem rotation

Figure 9.15 *Type SVB5 switchgear having horizontal isolation with cable earthing via top isolating contact stem rotation (courtesy of ALSTOM Medium Voltage Switchgear, South Africa)*

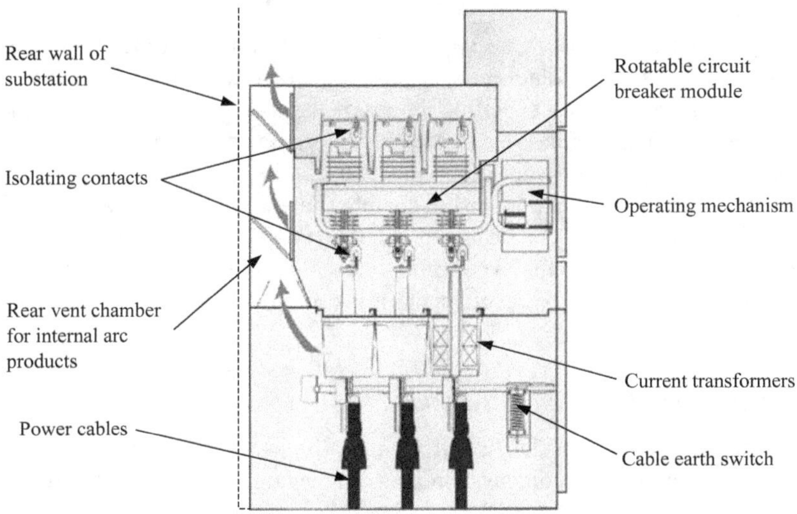

Figure 9.16 *The type VISAX circuit breaker. Isolation by circuit breaker rotation (courtesy of ALSTOM T&D Ltd)*

It will be seen in Figure 9.16 that the designer has provided access for making-off the power cables via the front of the unit, allowing the rear of the unit to be close to the substation wall and so saving space within the substation. In addition, the circuit breaker module, shown in Figure 9.17, has been designed to rotate through

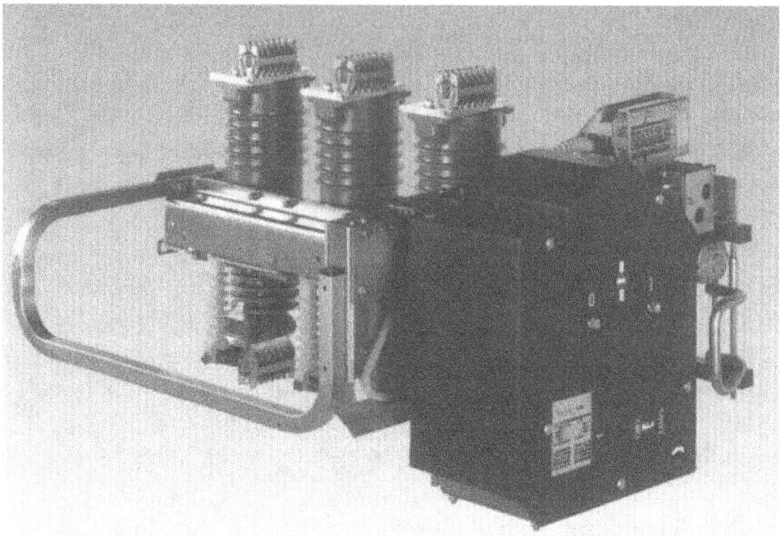

Figure 9.17 The type VISAX rotatable circuit breaker module (courtesy of ALSTOM T&D Ltd)

90° to provide isolation. Each phase post houses a vacuum interrupter, with isolating contacts at each end. The isolation of the circuit breaker takes place behind a closed front door. While this is not a specified requirement, it clearly will help to contribute to operator safety. The circuit breaker module incorporates runners, enabling it to be dismounted in the space provided for manual access.

The cable, circuit breaker and busbar chambers are arranged vertically, one above the other. This has allowed the designer to include a duct at the rear of the unit for venting of each chamber in the event of an internal arcing fault. The vent covers of each chamber are designed to prevent contamination from a faulted chamber entering an unfaulted chamber.

9.3.5 Gas-insulated primary switchgear

The SF_6 design technology developed for extra-high-voltage switchgear migrated with time down to medium voltage switchgear in the range up to 36 kV to provide the benefits that accrue from its use. These include size reduction, immunity from atmospheric contamination and also introduce the concept of maintenance-free switchgear. Figure 9.2, which shows a 33 kV duplicate busbar switchboard, the ALSTOM type WSB, serves to illustrate the dramatic savings in space that switchgear of this type can provide.

For simplicity and flexibility of production, switchgear of this type is manufactured in the form of modules that can be assembled in various configurations to meet the specific requirements of customers. This can be single or duplicate busbar units, bus couplers, bus section units, incoming feeder, outgoing feeder, units having

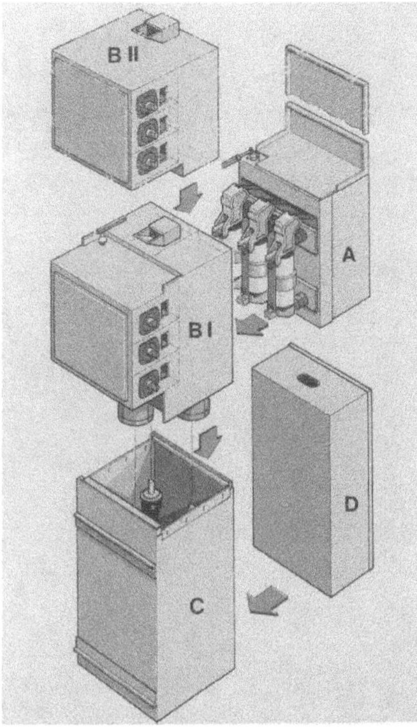

A Vacuum circuit breaker and mechanism.

BI Busbar chamber No.1

BII Busbar chamber No.2

C Sub-frame and cable chamber

D Low-voltage chamber

Figure 9.18 Type WSB 33 kV gas insulated switchgear (courtesy of ALSTOM T&D Ltd)

voltage transformer accommodation and so on. A diagram showing how a duplicate busbar GIS unit can be made up from individual modules is shown in Figure 9.18.

Gas-insulated switchgear has found favour in applications where space is at a premium or where the environment is particularly harsh, either because of high humidity, high salt levels, or where there is the possibility of dangerous industrial gases being present.

The switchgear designer has to make a careful choice with regard to the materials to be used within SF_6 gas. The gas is very dry and can change the mechanical characteristics of certain plastics that have a relatively high moisture content. It is also important that the switchgear is designed and proven to be discharge free internally. Even the smallest measure of discharge would cumulatively create dangerous acids that could lead to internal electrical failure.

Chapter 10

Cable connected secondary switchgear

It is the function of cable connected secondary switchgear to accept electrical power from a primary switchboard (see Chapter 9). The secondary switchgear then distributes the power to points in the network where the voltage is either transformed to a lower value or where it is consumed without transformation, as would be the case when supplying high-voltage machines.

There are basically two types of application. The first is that used to provide power to housing developments and small industrial estates and the second is within the local network of relatively large consumers. In the first type of application, a typical network used is the ring-main as shown in Figure 10.1.

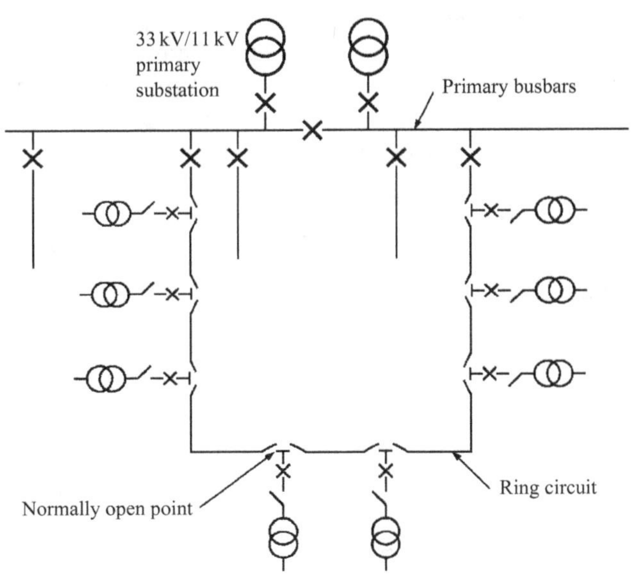

Figure 10.1 A typical ring-main network for secondary distribution

Within the typical distribution network shown, there are two primary incoming supplies, feeding the ring-main via two incoming circuit breakers and a normally open bus-section unit in the primary switchboard. The network is shown to be operating with one switch at the remote end open in the ring circuit. This means that each side of the ring will run as a radial feeder. This arrangement helps with fault location, reduces the number of consumers who would lose their supply in the event of a fault, giving the system greater flexibility and increased availability. In the event of a permanent fault occurring in one of the primary incomer circuits, the primary incoming circuit breaker would be opened/isolated and the bus-section circuit breaker closed to re-establish supplies to all consumers fed by the ring-main. Should a fault occur in any of the interconnecting cables, the ring switch on either side of the faulted section can be opened and the ring switch which was in the open position can be closed to re-establish supplies to all consumers. Repairs to the faulted cable section can then be carried out in safety with the ring switch at either end of the faulted cable opened and in the earthed position. Facilities must be provided to allow any cable repair to be tested before it is returned to service. Because of the large capacitance current taken by the cable, any a.c. voltage test sets would have to be very large, so it is normal practice to use a d.c. high-voltage test set for this purpose. The test voltages generally used for distribution cables and switchgear are shown in Table 10.1.

Figure 10.1 shows eight secondary substations being fed by the ring-main circuit, and in practice, up to a maximum of about 20 such substations can be supplied in this way. If we look in more detail at each secondary substation, we will see that they are all identical and have a common group of components (see Figure 10.2).

It was realised that as the same arrangement of distribution voltage switching functions was used at each T-off point, savings could be made by specifying factory built units that incorporated all of these functions within one standard unit. As will be seen in Figure 10.2, the combined functions that were incorporated into the new concept of ring-main unit, known universally as an RMU, were two ring switches, each having three positions of ON, OFF and EARTH, and a T-off switch to control the transformer, which was initially a high-voltage fuse-switch and later a circuit

Table 10.1 D.c. test voltages used for cables and switchgear

System rated voltage (kV)	15 min d.c. test voltage			
	New cables		Existing cables and switchgear	
	Phase–earth (kV)	Phase–phase (kV)	Phase–earth (kV)	Phase–phase (kV)
7.2	15	21	11	18
12	25	34	18	30
36	66	–	60	–

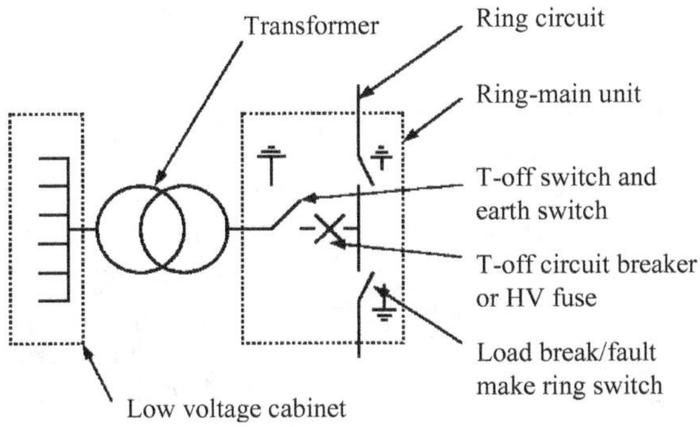

Figure 10.2 *Basic elements of a secondary ring-main substation*

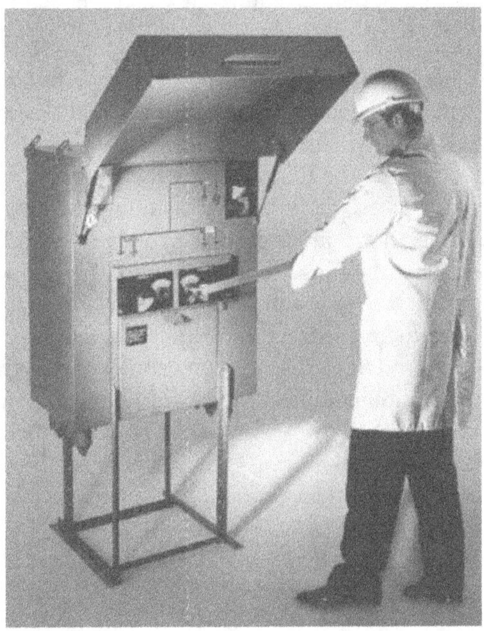

Figure 10.3 *A Schneider electric-type ringmaster SF_6 ring-main unit [15]*

breaker, together with a low rated transformer earthing switch. The ring switches were fitted with interlocked access covers to allow cable test plugs to be fitted to check the adjacent ring cable for possible faults.

The concept of the RMU originated in the UK but, because of the cost savings that they gave the user, the design and manufacture of RMUs was quickly taken up by manufacturers in many countries across the world. A typical modern outdoor ring-main unit is shown in Figure 10.3.

In addition to the direct cost savings associated with the use of a standard ring-main, additional benefits included:

(1) The purchaser could make an accurate direct comparison of prices between suppliers because of a standard specification.
(2) RMUs do not require a large R&D input when compared to circuit breakers. For example, the ring switches are usually of the plain break moving contact type, and the T-off circuit was originally almost exclusively controlled by high-voltage fuses. See the switching technology section of this chapter. These relatively basic levels of technology were readily available to a large number of manufacturers, allowing a greater number to compete, which had the effect of driving down the market price. In more recent times, the availability of low-cost, low rated vacuum interrupters and the development of rotating arc SF_6 interrupters have allowed larger transformers to be controlled although with a greater R&D effort and a marginal increase in unit costs.
(3) Site costs and space components required for interconnecting separate switching were saved.

The T-off connection between the ring-main unit is specified to be via a standard dimensional flange, which also has standard dimensioned electrical bushings. This gives the user interchangeability between various ring-main units as well as allowing connection to the associated power transformer either directly, or via a cable.

There were attempts to integrate further by including the distribution transformer in the same unit but these were unsuccessful. This was mainly due to the many manufacturers involved in combining components usually manufactured on different sites with these components requiring different manufacturing lead times. However, most manufacturers offer the equivalent to a skid mounted assembly of a ring-main unit, a transformer and low-voltage distribution cabinet (see Figure 10.4).

Although most distribution companies choose to directly couple the transformer and ring-main unit, some distribution companies prefer to mount each element of the secondary substation separately, using cables to connect them together. This is to allow for relatively easy replacement of individual elements. An example of this is shown in Figure 10.5.

The substation shown in Figure 10.5 is of the outdoor type. While substations of this type can minimise first cost, the equipment used has to be specified to have a high degree of environmental protection, typically IP54, with all ferrous parts being zinc plated. Even with this degree of protection, some problems can occur due to water ingress through seals that have to be made on site. The upper surface of the ring-main unit and its T-off cable box in Figure 10.5 can be seen to be suffering from the effects of the environment. The outdoor substation is also prone to external damage caused by vandals. The disadvantages of the outdoor open-type substation are overcome by the use of compact substations, housed within a cubicle that is factory built. An example of one of these is shown in Figure 10.6.

Figure 10.6 shows a secondary packaged substation with the HV and LV access doors open. Such arrangements offer the power distributor a small substation size with minimum installation work, together with good operator access. Visual impact

Figure 10.4 *Secondary substation with directly mounted ring-main unit (courtesy of Schneider Electric Ltd)*

Figure 10.5 *Secondary substation with cable connected elements (photo: author)*

Figure 10.6 Cubicle packaged secondary substation (courtesy of Schneider Electric Ltd)

Figure 10.7 Factory assembly of a part underground package substation (courtesy of ALSTOM Medium Voltage Switchgear, South Africa)

is becoming an important issue and, in order to minimise this impact, some manufacturers offer substation packages for installation either completely underground, or as shown in Figure 10.7, partially underground.

There are manufacturers who go to great lengths to offer equipment that will reduce the visual impact of substations. They do this using glass reinforced cement to manufacture the housings and applying an external finish that mimics other buildings in the area. These mimic the appearance of brick, stone and in extreme cases, even wooden log cabin-like constructions, as shown in Figure 10.8.

Returning to ring-main circuit applications. Some supply authorities, particularly those in southern Australia, have achieved even greater flexibility and availability by

Figure 10.8 *A glass reinforced concrete secondary substation in the Alps (courtesy of ALSTOM T&D Ltd)*

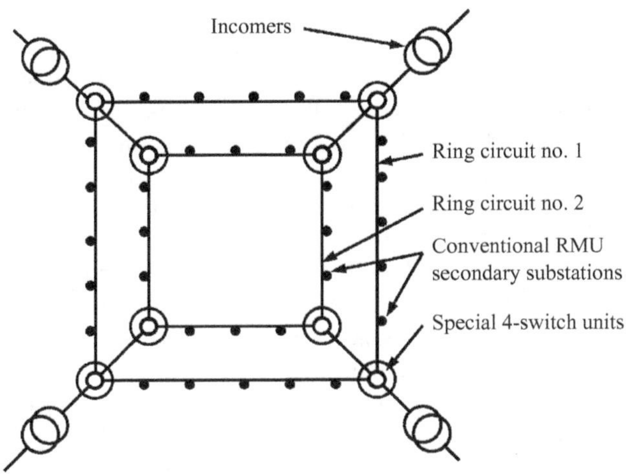

Figure 10.9 *An example of a concentric ring distribution network*

the use of concentric ring circuits in their secondary distribution circuits. An example of this type of circuit is illustrated in Figure 10.9, which shows a secondary distribution network having four incoming feeders. In practice, any number of feeders could be used. An arrangement of concentric ring-main circuits allows power to be shared between the ring circuits number 1 and 2 in the event of a circuit fault, resulting in a much greater degree of flexibility and availability.

It will be seen that the concentric ring network requires a four-switch unit at the interconnecting points of each feeder with the ring-main circuit. Units such as these were not readily available initially and required two standard ring-main units to be

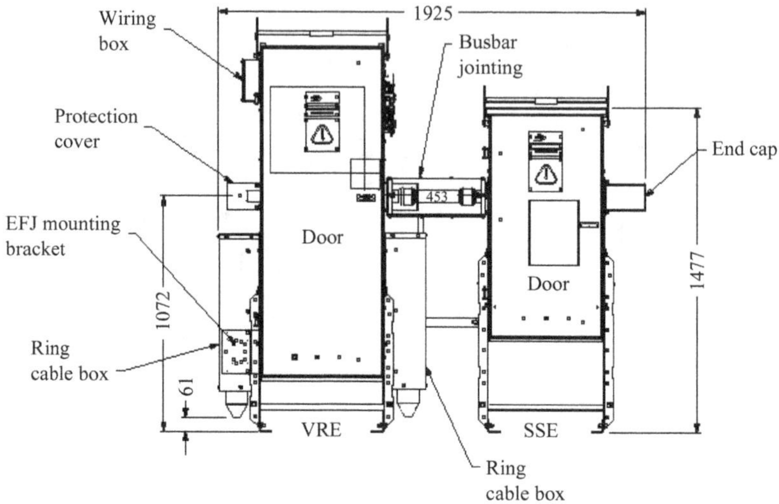

Figure 10.10 *Sabre extensible outdoor secondary switchgear (courtesy of W. Lucy & Co. Ltd)*

coupled together. However, continental manufacturers later developed four-switch units for applications such as these, but these new units were for indoor, or cubicle applications only. Within the United Kingdom, some manufacturers introduced extensible outdoor switchgear that would meet the requirements for concentric ring-main distribution schemes. A modern example of this type of switchgear is shown in Figure 10.10.

Some industrial users, particularly those involved in quarrying, have within their internal distribution system a need for two power transformers at each secondary substation. A number of these users saw that if a ring-main unit could be connected as a circuit spur via its T-off, cost savings could be realised by using one ring-main unit to control two power transformers via its ring switches, as shown in Figure 10.11. The users recognised that the T-off earth needed to be padlocked into the closed position as it only had a limited short-circuit rating. They also recognised that a fault in any transformer would result in both transformers losing their supply. However, for them this solution offered cost savings with an acceptable loss of flexibility and availability.

10.1 T-off circuit protection

The problem associated with the supply of power for the protection devices controlling the T-off circuit, which in turn supplies power to the distribution transformer, is that it is not readily available without the additional expense of a low-voltage supply. In this respect, it is very similar to the problems that needed to be resolved for the overhead line secondary switchgear (see Chapter 9).

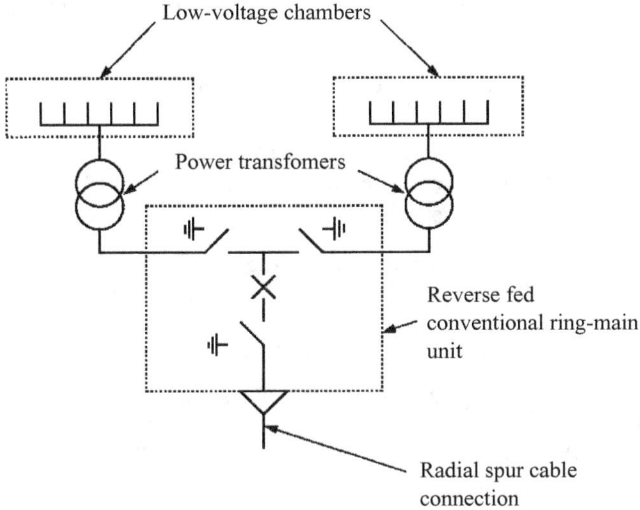

Figure 10.11 Reverse fed ring-main unit controlling two power transformers

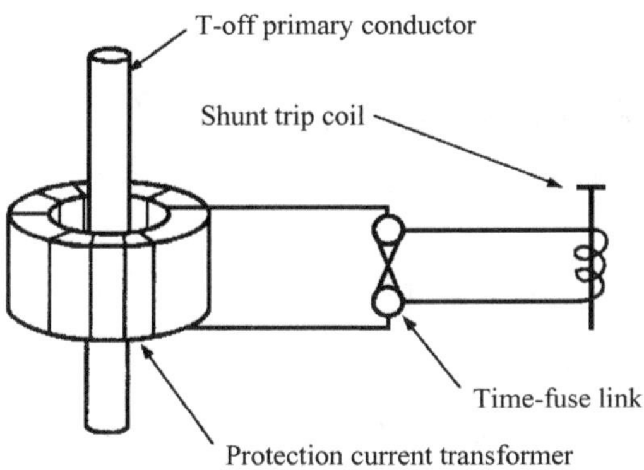

Figure 10.12 Time-fuse tripping circuit

Initially, when high-voltage fuse-links were used, the problem was resolved by fitting the fuse links with a small chemical charge which would trip all three phases in the event of any one fuse operating (see Chapter 12). With the introduction of circuit breakers for T-off control, they required an operating command to be given by the protection in order to trip under fault conditions. One of the ways of resolving this problem was to fit current transformers to the T-off bushings and arrange for each of these to be shunted by a special low-voltage fuse link, known as a time-fuse link. In the event that the current from the protection current transformer causes the time-fuse to operate, the output from the current transformer is automatically passed on to a shunt trip coil, which will operate to open the circuit breaker (see Figure 10.12).

Several factors have to be taken into account when selecting the rating of the time-fuse. These include:

(1) Catering for 150 per cent over-rating of the transformer, as allowed under the code of practice.
(2) The transformer inrush current, when being energised.
(3) Discrimination with the low-voltage fuses within the distribution cabinet.
(4) No tripping due to spillage current from the time-fuse circuit.

We examine each of these in turn using an 11 kV, 500 kVA transformer as an example.

10.1.1 *150 per cent transformer over-rating*

The normal full load current is given by the expression:

$$I_{\mathrm{fl}} = \frac{\text{kVA rating}}{\sqrt{3} \times \text{voltage rating}}$$

Given that the kVA rating is 500 and the voltage rating is 11 kV, the full load current will be:

$$I_{\mathrm{fl}} = \frac{500}{\sqrt{3} \times 11} = 26.24\ \mathrm{A}$$

The 150 per cent over-rating current will, therefore, be

$$I_{\mathrm{str}} = \frac{150 \times 26.24}{100} = 39.36\ \mathrm{A}$$

10.1.2 *The transformer inrush current*

The heating, or I^2t, of the transformer inrush current is conveniently taken as being equivalent to ten times the transformer full load current for 0.1 of a second. In other words, a point on the time current characteristic which should be less than that which would cause deterioration of controlling high voltage or time-fuse.

For a 500 kVA transformer, this would be 10×26.24 A, which gives the equivalent of 262.4 A at 0.1 s on the fuse operating characteristic.

10.1.3 *Discrimination with fuses on the low-voltage side*

Because of the distribution of fault current from the LV side of a transformer to the HV side, an inter-phase fault on the low-voltage side of the distribution transformer gives the most onerous condition in that this results in a fault current equal to 0.866 times its symmetrical fault current. The corresponding currents in the three phases on the high-voltage side of a Delta/Star connected distribution transformer are in the ratio of 2 : 1 : 1, the maximum being equal to the high-voltage side three-phase fault current as against 0.866 times the symmetrical fault current on the low-voltage side. Given that the fault impedance of the distribution transformer is typically 5 per cent, and assuming that a factor of 0.6 is used to take into account the low-voltage impedance of the fault. The fault current which will flow on the high-voltage side will be equal to:

$$\frac{100 \times 0.6}{\sqrt{3} \times 5} = 6.93 \text{ times the rated high-voltage current.}$$

which, for a 500 kVA, 11 kV transformer:

$$= 26.24 \times 6.93$$

$$= 181\ \text{A}.$$

Given the protection current transformer ratio, and the distribution transformer ratio, the current on the secondary side can be determined. This latter ratio is equal to the voltage ratio multiplied by 0.866. For example, a 11 kV to 415 V transformer will have a transformation ratio of

$$= \frac{11{,}000 \times 0.866}{415} = 22.95.$$

With this information, the operating time for the low-voltage fuse can be compared with the operating time of the protection time-fuse to determine compatibility.

10.1.4 No tripping due to spillage current from the time-fuse circuit

Care has to be taken to ensure that under maximum fault conditions, the impedance of the time-fuse circuit is not such that spillage current from this circuit be sufficient to operate the circuit breaker trip coil. To this end, the time-fuse circuit, complete with fuse should be tested under voltage and current conditions which are considered to be typical with regard to waveshape, with a magnitude equal to or greater than that which would be found in service.

Standardisation of the protection current transformer ratio has been introduced and these ratios together with time-fuse ratings are shown in Table 10.2, which follows the time-fuse characteristics.

10.2 Time-fuse operating characteristics

Figure 10.13 shows time-fuse characteristics that have become standard within the UK.

It will be seen that not all of the time-fuses shown in Figure 10.13 are included in Table 10.2. Those not shown are intended for use with non-standard transformer ratios and non-standard protection CT ratios.

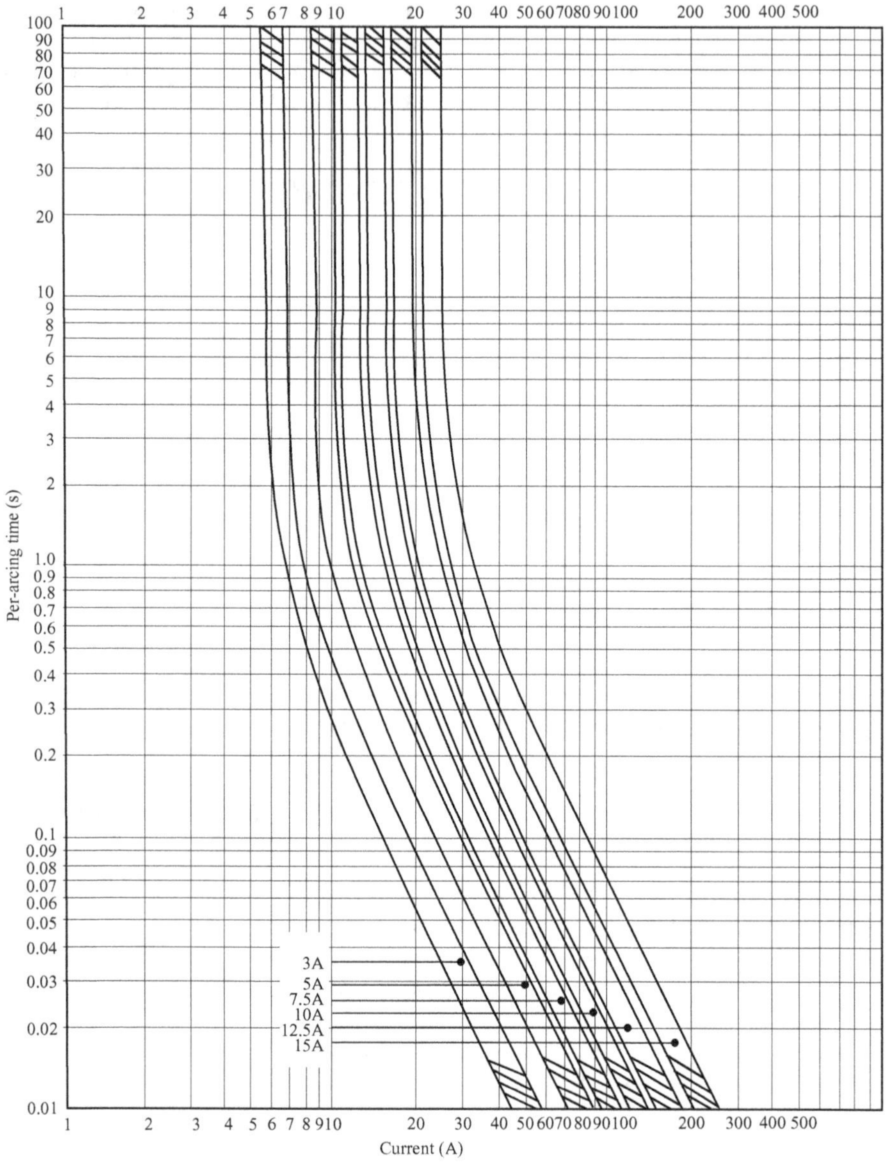

Figure 10.13 English Electric time-fuse operating characteristics (courtesy of ALSTOM T&D Ltd)

10.3 The Falcon protection scheme

The Falcon ring-main unit was the first SF_6 ring-main unit having a circuit breaker to control the T-off circuit and was introduced by Hawker Siddeley in the early 1980s.

Table 10.2 Protection CT ratios and time-fuse ratings

Transformer		Protection CT ratio	Time-fuse rating (A)
Voltage (kV)	kVA rating		
11	315	50/5	5
11	500	50/5	10
11	800	100/5	7.5
11	1000	100/5	10
6.6	315	50/5	10
6.6	500	100/5	7.5
6.6	800	100/5	12.5

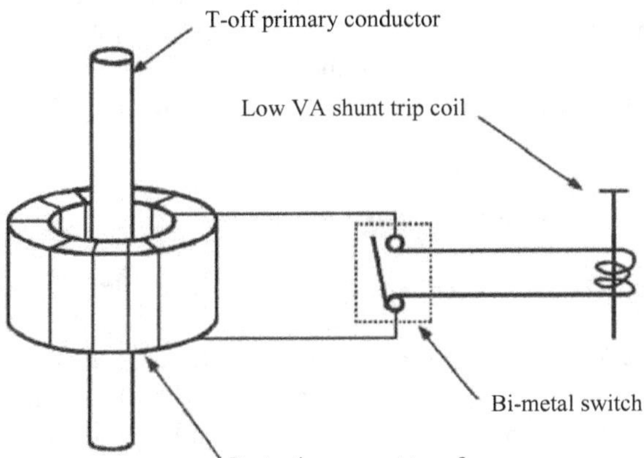

Figure 10.14 The Falcon type protection

The design philosophy that was adopted was that no fuses, either high-voltage or time-fuses would be used as these would have to be stocked and carried by operational personnel, which was considered by the designers to be undesirable to the potential users. A means had, therefore, to be found to detect fault conditions and to trip the T-off circuit breaker when pre-set limits were reached, without any low-voltage supply being available. In addition, this protection had to be sufficiently flexible to be applied to the protection of a wide range of distribution transformers.

The solution, which was arrived at and patented by Hawker Siddeley, was similar to that used in the time-fuse system but with the time-fuse replaced by a bi-metal switch, which was supplied by very small current transformers. These current transformers had cores that were designed to saturate and, in conjunction with a range of bi-metal element switches, produce a family of protection curves which meet the criteria set out for the time-fuse type protection. Figure 10.14 illustrates the protection arrangement.

As the current transformers used in the Falcon protection scheme were relatively small, the power output for tripping the T-off circuit breaker was also very small. This meant that a low VA trip coil had to be used with a rectifier interposed between it and the bi-metal switch.

10.4 Protection where a low-voltage source is available

The provision of a low-voltage source local to secondary switchgear is relatively expensive, but in certain circumstances it can be considered to be essential. Such circumstances include applications requiring facilities for remote measurement, operation and control. The benefits of these facilities include a reduction in the number of customer minutes lost in the event of a system fault; one of the criteria by which the efficiency of local electrical distribution companies is judged. However, once a local low-voltage electrical supply is available, more sophisticated protection and control equipment may be employed. An example of a remote terminal unit, known as an RTU, for communication between the ring-main unit and its control centre for monitoring conditions at the site, is shown in Figure 10.15.

10.5 Secondary distribution switchboards

In large customer networks, there is a need for secondary distribution switchboards. These switchboards take their power from the network via the supplier's primary incoming switchboard. Secondary switchgear used for these distribution applications usually has lower fault and normal current ratings and less flexibility than switchgear used for primary distribution, and in addition, it may have reduced dimensions. For

Figure 10.15 A Talus 200E remote terminal unit for secondary distribution (courtesy of Schneider Electric)

example, the switchgear used in secondary distribution switchboards will usually not be suitable for applications requiring duplicate busbars and quite often the circuit breakers will be of the fixed type as opposed to the withdrawable type most common in primary distribution switchgear. However, as these switchboards are normally of the multi-panel type, they will usually be provided with a low-voltage power source, permitting more sophisticated measurement, protection and control than would normally be found in ring-main units. The circuit breaking elements of these units are often derived from those that were developed for secondary overhead line and ring-main unit distribution applications. This can be seen in both the size and ratings of vacuum interrupters and in SF_6 interrupters. An example of this with SF_6 interrupters can be seen in Figure 10.16(a) and (b).

(*a*) (*b*)

Figure 10.16 *a* The interrupter enclosure of the Ringmaster RMU [3]
b A sectional view of the GENIE circuit breaker (courtesy of Schneider Electric)

(*a*)

(*b*)

(*c*)

Figure 10.17 *a* A Ringmaster extensible switchboard (courtesy of Schneider Electric)

b Indoor secondary switchgear within a cubicle for outdoor application (courtesy of ALSTOM Medium Voltage Switchgear, South Africa)

c Type VISAX 'S' fixed vacuum switchgear (courtesy of ALSTOM T&D Ltd)

Most manufacturers produce secondary distribution switchgear using either vacuum or SF_6 interrupting technology. The primary insulation systems can be air, SF_6 gas, or SF_6 gas with solid insulation as shown in Figure 10.16(b).

The design of the switchgear can be of the indoor or outdoor type, or even indoor design within a cubicle type housing for outdoor applications. An example of an outdoor switchboard is shown in Figure 10.17(a). Examples of some of the other types are shown in Figure 10.17(b) and (c).

Chapter 11

Overhead conductor connected secondary switchgear

11.1 Introduction

Because of the relatively high cost of electrical distribution by buried cable, power supplies to sparsely populated areas are invariably provided by overhead line conductors. The present-day overhead line distribution equipment is the product of innovation and evolution resulting from many years of application in the field of rural electrical distribution. Advances in technology, as they have become available, have been adopted to provide better security and continuity of the supply of power to remote centres of consumption, as well as providing the power supplier with information which was not readily available a few years ago.

11.2 Standards

The basic standard for autoreclosers is ANSI C37.60. This standard was modified when vacuum interrupters were introduced to recognise the longer contact life of the new technology. The standard requires many more fault operations with vacuum than with oil, recognising the inherent longer life of vacuum interrupters.

The only other known standard for autoreclosers is the UK EATS41-26, which accepts the ANSI standard but adds to it the IEC values for transient recovery voltage as well as introducing a three second short time current fault rating and internal arc fault requirements.

11.3 Historical background

Electrical power supplies to remote consumers are invariably provided by radial overhead line feeders with pole mounted transformers at the consumer's end to provide a

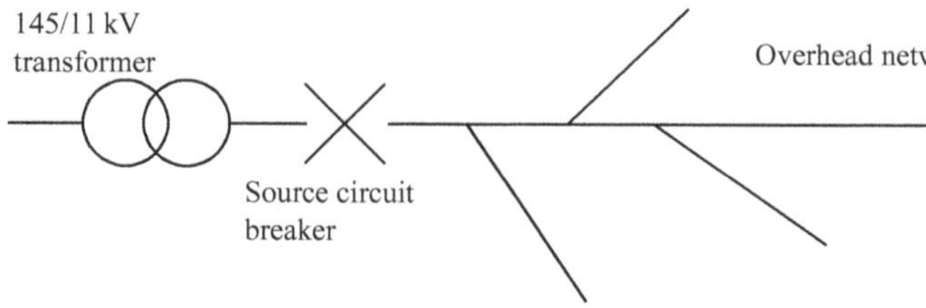

Figure 11.1 A typical early rural network with source circuit breaker only

low-voltage supply. The voltage of the overhead line from the source to the consumer's transformer can be anything from, typically, 3.3 to 38 kV.

The earliest systems consisted of a source transformer supplied at a voltage between 66 and 145 kV, which stepped this voltage down to 3.3 to 38 kV, and supplied the overhead line via a source circuit breaker (see Figure 11.1).

When a fault on the line caused the source circuit breaker to trip, all consumers would lose their electrical power, and a patrol would set out on foot to find and repair the fault. Experience showed that on many occasions no fault could be found, as the fault was transient in nature, caused by birds, lightning, wind blown foliage and the like, which meant that the source circuit breaker could be successfully reclosed, re-establishing power to the consumers. Human nature, being what it is, meant that operators would 'test' the line by closing the source circuit breaker before deciding to set out to find the fault, and in more than 50 per cent of the cases, supply was successfully restored without line inspection being necessary. This was extended by the operators attempting a second closure of the source circuit breaker, should the first attempt result in a trip operation. This was found to result in about another 25–30 per cent successful re-establishments of supply. It was correctly deduced that most faults on overhead power lines were transient in their nature, the fault being self-cleared when the electrical power was removed.

In the early 1930s, it was decided that the immediate need was for a new operating mechanism on the source circuit breaker which would automatically reclose the circuit breaker three times when it tripped out on a fault (see Figure 11.2).

This technique reduced the time necessary to restore supplies to customers when a transient fault appeared on the system. The application was successful in that power was restored when interruptions were caused by transient faults, but all customers lost power for a short period regardless of where the fault occurred within the system.

11.4 Pole mounted autoreclosers

It was realised that if a circuit breaker, preferably pole mounted, could be installed on the overhead line some distance into the network to automatically open and reclose in the event of a fault, the disadvantage of all customers losing their electrical power whenever a fault appeared would be overcome, by avoiding the interruption of supply to customers connected upstream from the breaker. The problem facing

Figure 11.2 *An early source circuit breaker fitted with a four shot to lock-out mechanism, circa 1931 (courtesy ALSTOM T&D Ltd)*

the switchgear designer in producing such a device was how to provide the operating mechanism energy. It was found that this operating energy could be provided by a weight suspended from the operating mechanism to recharge the mechanism springs. Autoreclosers having this type of operating mechanism were then produced and installed. They brought a significant improvement to the system in terms of what is today called 'customer minutes lost' as customers upstream from the autorecloser saw no interruption in supply and customers downstream from the autorecloser saw only a short interruption if the fault was transient in nature. The disadvantage of this arrangement was that for practical height reasons energy for only six operations could be stored before the weight reached its bottom limit, which required a visit from an operator to haul the weight back to its top position. Statistics showed that a pole mounted autorecloser installed on an average length overhead line in an average area would operate up to 138 times in a year. Weight operating mechanisms, while bringing improvements to the system, still involved considerable manual intervention to keep them operational, requiring 20–30 visits to site per year per unit, to haul the weight back to its top position. An autorecloser of this type is shown in Figure 11.3.

It was an American, by the name of Kyle, who, in 1956, introduced the use of power from the overhead line itself to recharge the mechanism springs. He did this by including an HV solenoid within the autorecloser tank that was automatically connected to the overhead line each time the unit was opened, until a pre-set number of operations in a given time had been reached. A schematic for this type of unit is shown in Figure 11.4. This shows that the power supply for the HV solenoid is provided

Figure 11.3 *A weight operated pole mounted autorecloser (courtesy of ALSTOM T&D Ltd)*

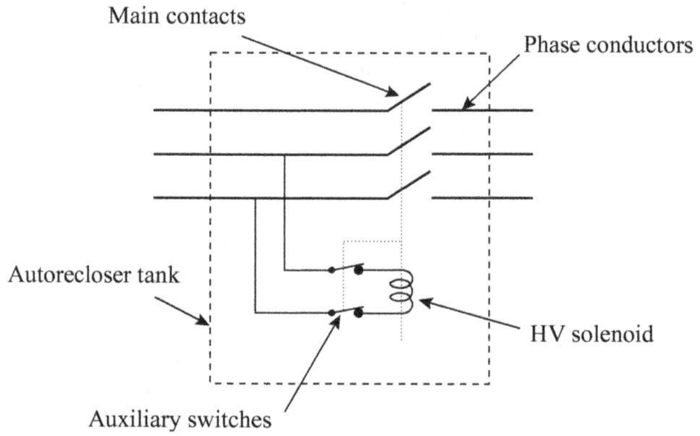

Figure 11.4 *Schematic of an autorecloser operated by an internal HV solenoid*

by electrical connections between two phases on the supply side of the autorecloser. The supply is controlled by two auxiliary switches that are automatically closed when the main contacts move to the open position, thus connecting the HV solenoid to the supply voltage and causing the unit to close. Not shown in the schematic is another pair of series connected auxiliary switches which, when opened, will isolate the HV solenoid leaving the autorecloser in the 'locked out' condition. These additional auxiliary switches are operated by a device that causes them to operate when a pre-set number of operations occur within a defined period of time.

11.5 Technical terms

Before discussing the application of these and subsequent units on a distribution system, it is necessary to appreciate the meaning of a number of special technical terms, which are in addition to those applied to normal circuit breakers:

Dead time
This is the elapsed time from the cessation of current flow to its re-establishment.

Shots to lock-out
The pre-set number of fault trip operations before lock-out.

Lock-out
The open condition reached by an autorecloser when it has completed its pre-set number of shots to lock-out. Manual local or remote intervention will be necessary to close the unit.

Delayed trip
A trip operation with a time delay inversely proportional to the magnitude of the fault current.

Instantaneous trip
A trip operation which is performed without any built-in time delay.

Sequence
The pre-set trip operations within the shots to lock-out. These will be either delayed (D) or instantaneous (I) trip operations. The most commonly specified sequence was 2I 2D. The reasoning behind this was that the two instantaneous trips would clear any transient system faults, and after those operations the fault, if it was still present, was deemed to be permanent and the delayed operation was to allow sufficient energy to pass through downstream fuses to clear the fault (see Figure 11.5).

Reclaim time
The time taken by the autorecloser's protection to reset to the beginning of its pre-set sequence should the fault be cleared before lock-out is reached.

It will be seen in the upper part of Figure 11.5 that the transient fault current, represented by a relatively thick line, causes the autorecloser to trip on an instantaneous setting. On the subsequent reclosure, the fault is still present, causing the autorecloser to again trip, this time on its second instantaneous setting.

On the third reclosure, the transient fault has cleared, the autorecloser remains closed and the protection resets in readiness for any subsequent fault. In this case, all downstream customers have their power supply restored.

The lower part of Figure 11.5 shows the sequence of events when a permanent fault occurs on a system having an autorecloser and downstream high-voltage fuses.

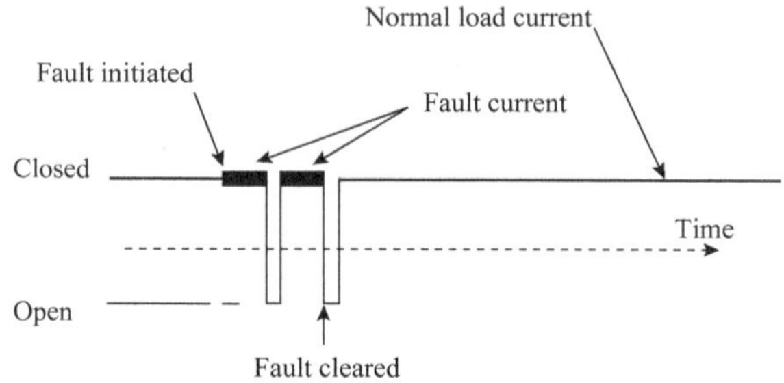

Transient fault cleared by the second instantaneous autorecloser trip

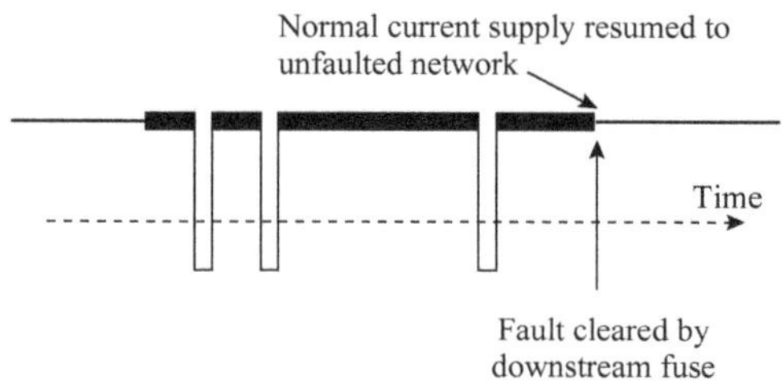

Permanent fault cleared by a downstream fuse during second delay trip

Figure 11.5 *Autorecloser sequences when clearing transient and permanent faults*

The fault causes the autorecloser to carry out its two instantaneous trip operations and then, as the fault is still present, initiate its delayed trip sequence in order to allow sufficient fault current to flow to cause the downstream high-voltage fuses to operate and clear the fault. The autorecloser remains closed and its protection resets in readiness for any subsequent fault. In this case, all customers downstream of the high-voltage fuses will lose power until repairs are carried out, and all other customers will have their power restored.

11.6 Discussion on autoreclosers with HV internal solenoid mechanisms

While the introduction of autoreclosers with internal HV solenoids brought an improvement in the continuity and reliability of the supply of power to rural consumers, they also made life difficult for the manufacturers. This was because of a natural law associated with solenoid operation, which limits the voltage range to which they can be applied. This range is about ± 10 per cent. Too low an applied voltage would result in prolonged coil energisation, leading to burn out, bursting, and internal fault. Too high a voltage would cause mechanical damage to the mechanism. As the actual specified system voltage can be anything between 3.3 and 38 kV, it meant that an individual customer's system voltage had to be identified at an early stage in a contract, and, therefore, manufacture for stock was not practical. It also meant that routine test equipment within the factory before dispatch had to be sufficiently flexible to cope with the operation of the units at all system voltages, as these tests checked the operation of the autoreclosers on their pre-set sequences at the specified system voltage. It also meant that customers needed a high-voltage test cage within their maintenance depots to check new units on arrival and units which were being maintained. Finally, because of the electrical connections of the internal high-voltage solenoid, the autoreclosers could only operate with the power source in one direction, effectively limiting their application to radial connected overhead lines effectively precluding ring circuit applications.

11.7 Hydraulic control

Until about 1980, most autoreclosers used hydraulics to control the type of trip operation, I or D, and the reclaim time of the protection. Phase fault trip operations were via series connected coils, which were relatively small solenoid coils within the autorecloser, one per phase, and were connected in series with the main current circuit of the overhead line. The series coil assembly included hydraulic dashpots connected to armatures located within each solenoid. In the event of a system fault, the armature would be attracted into the solenoid by the passage of fault current. For instantaneous trip operations, a port in the side of the dashpot would be opened, and for delayed operations the port would be closed, bringing the dashpot into action. It will be appreciated that each current rating would need its own size of series connected coil, in order to maintain constant ampere turns, further limiting the locations on the system where the autorecloser could be installed without internal modification. The selection of I and D trip operations and the number of shots to lock-out could only be adjusted by the position of drive pins within the operating mechanism, making such changes practically impossible in the field. The reclaim time for the protection could not be adjusted, as this was set by a dashpot that had to be manufactured to a very high accuracy to obtain limits of 60–120 s. It should be noted that time control using hydraulic means would vary with temperature due to changes in oil viscosity, which could lead to protection discrimination problems.

The dead time of the autorecloser was determined by the natural spring charging time for the operating mechanism, typically 1 s. Later models were provided with a mechanism drive point adjustment which could double the dead time. No other dead times could be selected. Despite all of these disadvantages, units of this type were successful because of the benefits they provided. It is interesting to note that the identified requirement for earth fault protection was met by using three current transformers and a spring loaded relay. Current transformers were not normally used for phase fault protection until electronic control relays were introduced in the 1980s.

11.8 The short-circuit fault level of overhead lines

Because of the relatively high impedance of overhead lines, there is a rapid fall off in fault level with the line length from the transformer source. For example, a 14 kA fault level at the source will be reduced to only 4 kA at about 4 km from that source.

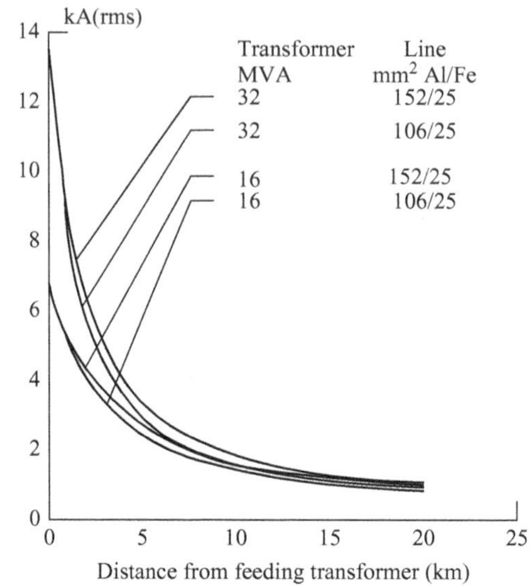

Figure 11.6 Fall-off of fault level with line length on a 13.8 kV system [16]

This is illustrated in Figure 11.6 which gives values for a 13.8 kV system. As overhead lines can be 100 km long, the fault level at the mid-point will be very low. For this reason, 5 kA was a typical standard rating for an oil autorecloser.

11.9 Advances in interrupter technology

In the early 1980s 'sealed for life' autoreclosers using SF_6 rotating arc interrupters were introduced which eliminated the three-year maintenance requirement of the oil interrupter autoreclosers. The contact drive system of a unit of this type is shown in Figure 11.7. A more detailed explanation of the way in which rotating arc interrupters function is given in Chapter 2. In many ways, the drive system for an autorecloser using rotating arc SF_6 for interruption is very similar to the oil interruption types and shares the common loss of flexibility in application with regards to the direction of power supply.

However, there were two fundamental advantages offered by this new design. The first was that during fault current interruption, the rotating arc SF_6 interrupter did not impose any significant back-pressure on the contact system, thus reducing total mechanism energy necessary for operation. Secondly, and more importantly, for the first time, a unit requiring no contact maintenance was being offered. The many advantages of using SF_6 for both interruption and insulation are detailed in Chapter 2. However, there were two significant problems that caused some delay in the subsequent rapid acceptance of this type of unit. The first was the customers'

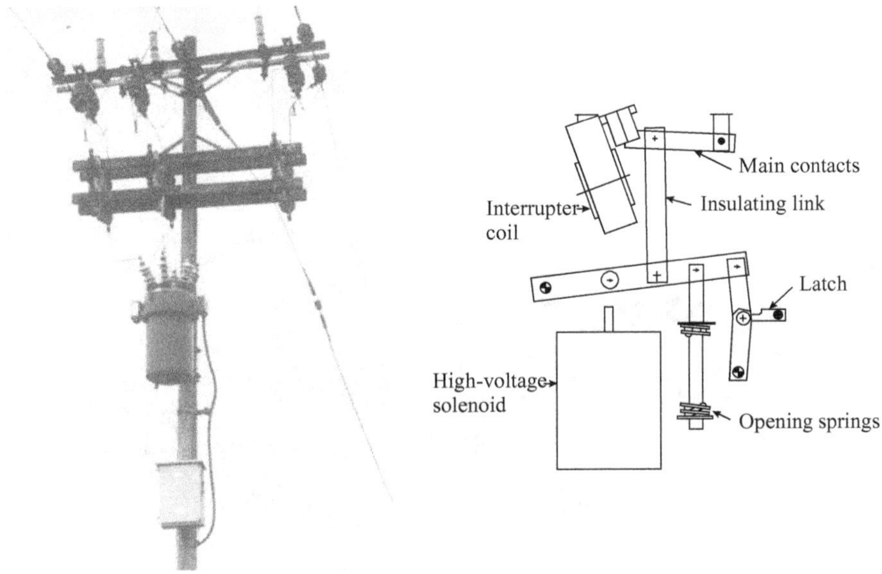

Figure 11.7 *The type PMR unit. An* SF_6 *rotating arc autorecloser [17]*

need to gain confidence in gas tightness and the second, which was shared by all SF_6 switchgear, was concern about end of life disposal.

At about the same time as the SF_6 autoreclosers were being introduced, the Kyle plant in the USA introduced oil immersed vacuum interrupter autoreclosers. Both the SF_6 rotating arc and vacuum interrupter types of units were capable of operating at fault levels up to 16 kA. This made them very attractive to customers for use within the system and also, because of their ratings, as pole mounted source circuit breakers in place of the much more expensive conventional ground mounted circuit breakers.

11.10 Sectionalisers

Overhead line sectionalisers are fault make/load break switching devices which are designed to automatically open during the dead-time of an associated upstream autorecloser, in order to remove a faulted section of the line and allow the power supply to be restored to the remainder. Sectionalisers operate by counting the number of pulses of fault current within a specified period of time and only opening when the passage of fault current ceases. The settings are adjustable to allow up to three pulses of fault current to pass before operation. These pulses of current can be of very short duration, such as would be the case if the autorecloser was set for instantaneous trip operations. Therefore, sectionalisers extend the benefits of autoreclosing further into the network. A diagram illustrating the co-ordination between a sectionaliser, set for two current pulses to trip, and its upstream autorecloser is shown in Figure 11.8.

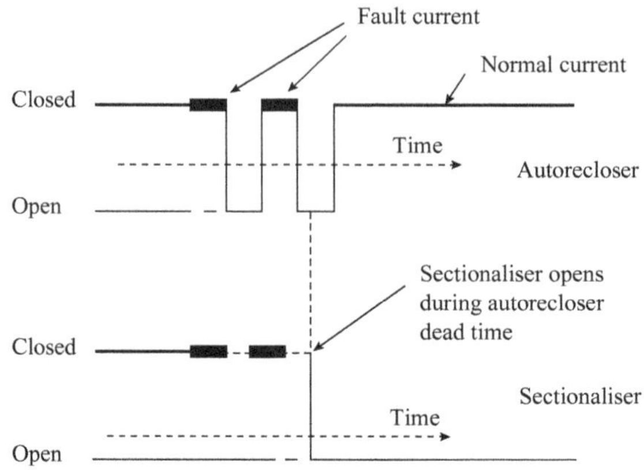

Permanent fault cleared by a downstream sectionaliser

Figure 11.8 Autorecloser and sectionaliser co-ordination

As sectionalisers can be fitted with earth fault protection, they can detect and help to remove earth faults with currents as low as 4 A. High-voltage fuses, on the other hand, cannot detect these low currents and, therefore, do not have the same functionality. As sectionalisers are only load-break units, and the energy for opening their contacts is stored in springs within the unit, there is a potential for danger. They are the only load-break devices in distribution systems that are held closed by a mechanism latch. Manufacturers of sectionalisers recognise that the integrity of the latch is of the utmost importance as it would be catastrophic if, during the passage of fault current, the latch were to be released by shock or vibration.

Examples of typical single-phase and three-phase sectionalisers are shown in Figure 11.9(a)–(c).

There is another type of sectionaliser available that is known as an 'automatic sectionalising link'. These are factory-sealed devices that are a direct fitting replacement for distribution type expulsion fuses. They consist of a sealed single metal tube, which acts as a Faraday cage, enclosing an electronics package. Surrounding the tube is a small current transformer that inputs data to the electronics relating to the current flow. If the threshold of operating current is exceeded, that is when fault current flows, the electronics will count this as fault number one.

If the fault is transient in nature and is cleared by the associated upstream autorecloser, the electronics will reset to zero. However, if the fault current resumes when the autorecloser closes, the electronics, if factory set for two pulses to lock out, will wait for the fault current to cease for the second time, and then send an ignition current to a chemical striker pin. This, in turn, causes the automatic sectioning link to drop out to the 'open' position. These links can be arranged to cause all three phases to open if one phase operates. The device is shown in Figure 11.10.

The automatic sectionalising link has the distinct advantage over more conventional sectionalisers in terms of cost, but the current transformers are too small to

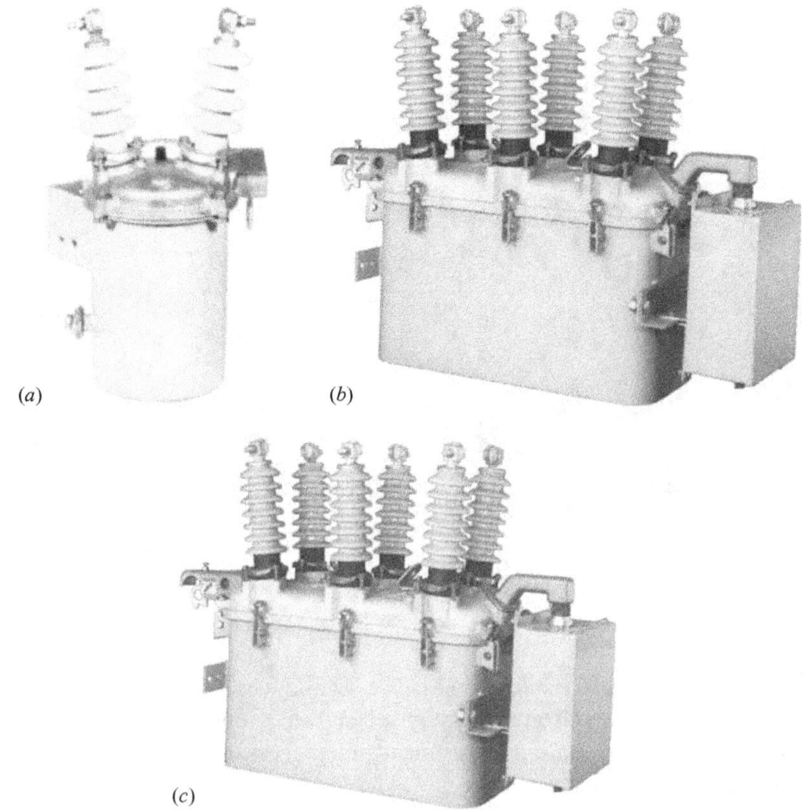

Figure 11.9 *Examples of sectionaliser units*
a Single-phase type GH 14.4 kV, 140 A
b Type GN3E 14.4 kV, 200 A, and
c Type GN3VE 24.9 kV 200 A with electronic control
(courtesy of Cooper Power Systems)

be residually connected so they have the disadvantage of not being able to see and operate on earth fault currents.

As autoreclosers can be used in series, both of the above types of sectionalisers owe their existence to their relatively low unit cost when compared with the unit cost of an autorecloser.

11.11 Protection

At about the time that the new interrupting technologies were being introduced, work was being carried out to produce a microprocessor protection relay to control all of the functions that were currently being carried out by hydraulic means. There were several objectives to carrying out this work. The first was to reduce the mechanical

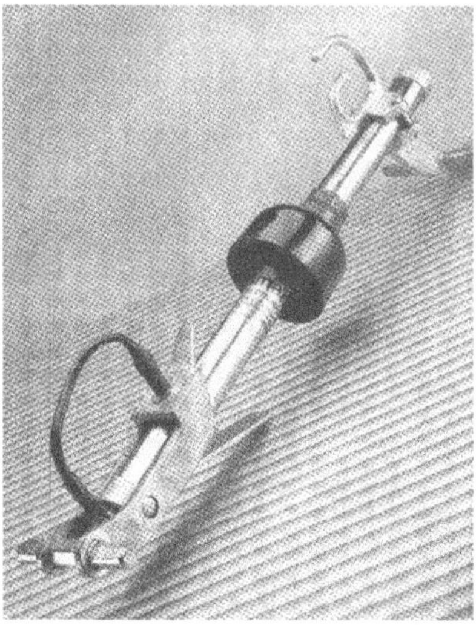

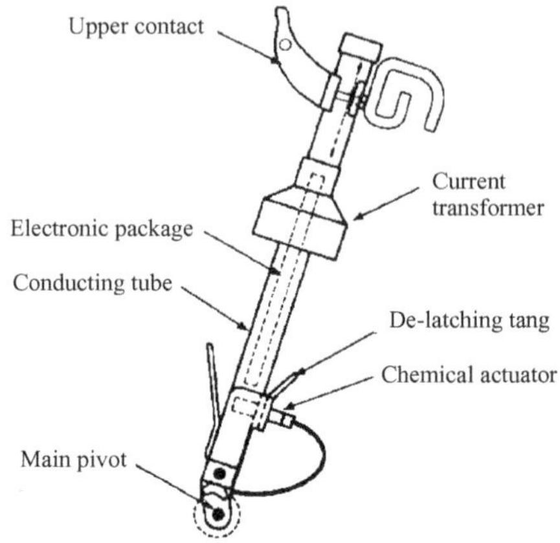

Figure 11.10 The automatic sectionalising link (courtesy of Cooper Bussmann of Cooper (UK) Ltd)

complexity of the hydraulic control types. The second was to give greater flexibility in application, with a much greater range of customer options. The third was to reduce the man-hours in test and adjustment time necessary with hydraulic control that could be up to one man week per unit. The fourth was to eliminate variations in settings caused by the effect of temperature on the oil viscosity, which has a marked influence upon operating and reclaim times and protection curves. The final objective was to provide additional features that were perceived to be desirable, such as a large selection of protection curves, fault record history and remote control interfaces to meet the growing demand for system automation. The use of microprocessors has meant that, in many ways, only the imagination sets a limit on the facilities that can be provided. A problem facing the relay designer was the same one as was previously solved by the switchgear designer, that was the lack of a local low-voltage source to power the relay. One solution, which is currently employed by some manufacturers, is to take the relay power supply from a local low-power voltage transformer. At about this time, long-life lithium batteries were becoming available. However, the current taken by the relay, about 40 mA, was too high to achieve a realistic battery life. A solution to this problem was found by arranging for the relay to power-up only when required by the control or protection, which was achieved by powering comparator circuits within the relay. This meant that the power consumption of the relay was reduced from 40 mA to only 100 μA when in quiescent mode, allowing the battery solution for relay power to be adopted. However, because of concerns about battery life and reliability, there is a growing trend towards the use of rechargeable batteries, which receive their charge supply from voltage transformers on one or both sides of the autorecloser.

11.12 Magnetic actuators and their impact on the design of autoreclosers

Up to this stage, the progress of autorecloser development had reached the point where ratings had increased to those of source breakers, but they still took power from the overhead line for operation. Therefore, although they had established electronic control and protection relays with their own independent lithium battery supply, the problem of dependence upon power from the overhead line for operation still remained. The relatively low operating energy requirements for vacuum interrupters, together with the advent of new rare earth permanent magnets, allowed switchgear designers to examine an idea for a mechanism drive, based upon the switching of magnetic flux. Chapter 8 discusses the design of magnetic actuators in greater depth, but to gain an understanding of the principle of their operation, see Figure 11.11.

The assembly shown in Figure 11.11 is a magnetic actuator in its simplest form. It has a single moving part, the central armature. This armature has a ring of permanent magnets assembled within the fixed structure at the mid-point of the armature. Above and below the permanent magnets are two coils. When the upper coil is energised, the magnetic flux that it produces assists the permanent magnet flux in the air gap above the armature, and opposes the permanent magnet flux which is holding the armature in the lower position. The value of flux density is chosen to be such that the armature will be attracted to the upper position and will then be held there by the permanent magnets when the coil is de-energised. The reverse occurs when the lower coil is energised. The design is therefore inherently bi-stable. The combination of magnetic actuator and vacuum interrupters resulted in a mechanically simple device, and opened up the opportunity for autoreclosers having lithium battery operation, thus regaining the independence from the overhead line for operation that was lost when the high-voltage solenoid was introduced.

The advantages for the user of magnetic actuator mechanisms include the flexibility of application at any system voltage up to the rated voltage without modification, and the possibility to operate with power sources in either direction relative to the

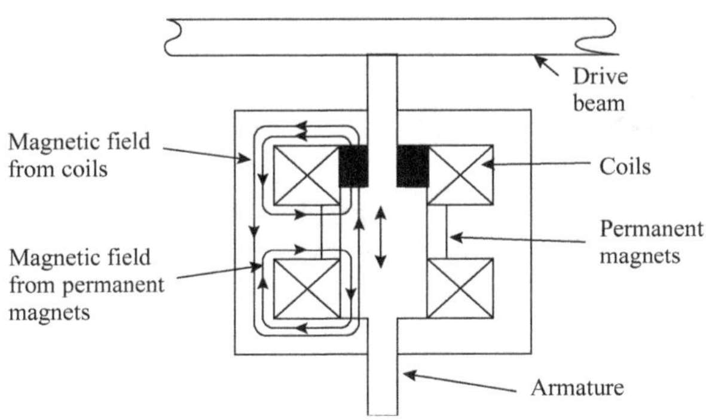

Figure 11.11 Schematic diagram of a magnetic actuator

autorecloser's orientation. The magnetic actuator also gave the advantage of obviating the need for a central high-voltage test cage to check autorecloser operation.

For the manufacturer, the advantages were that in addition to mechanical simplicity, it was no longer necessary to manufacture units to specific customer system voltages. This meant that units no longer had to be manufactured in small batch requirements, and that in-house operational checks could be carried out without the need for a high-voltage test cage. However, the advantages for both the customer and manufacturer did not end there. As unit operation was, once again, independent of the overhead line, the direction of the power source relative to the autorecloser did not matter, giving considerable flexibility in application.

11.13 Remote monitoring and operation

There is a growing demand for remote monitoring and operation of autoreclosers. Remote monitoring gives the supply authority knowledge of the state of the network; which switching units are closed or open, the current flowing, without having to wait for notification of a problem from the public before taking corrective action. Remote operation provides the opportunity to reconfigure the network in order to restore supplies as soon as possible in the event of a permanent system fault.

11.14 Islands of intelligence

This system, which was devised by power supply companies on the north east seaboard of the USA, provides automatic network reconfiguration without the use of radio or any manual intervention or inter-unit communication other than the passage of fault current. It is known as a closed loop system. A diagram of this type of system is shown in Figure 11.12.

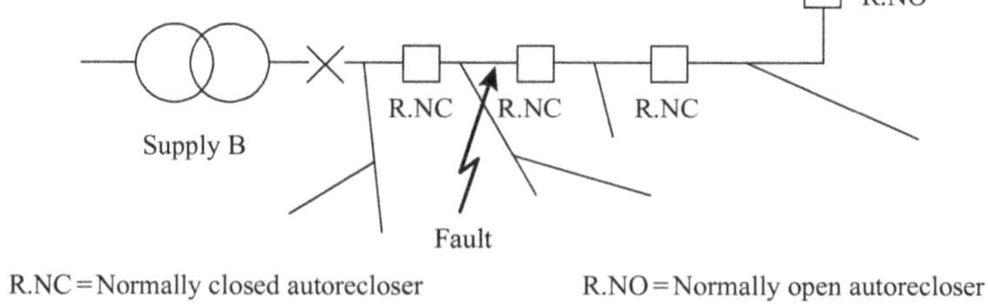

Figure 11.12 Diagram of a closed loop system

Consider Figure 11.12. It can be seen that the system consisted of two radial feeders, A and B, each having three autoreclosers. Another autorecloser, normally open, is connected between the two lines at their remote ends. Each autorecloser had voltage detection on each side. It can be seen that if a permanent fault occurred beyond the first autorecloser on line B, the second autorecloser on this line would detect loss of voltage on one side and would itself open when a pre-set time was reached. The normally open tie autorecloser would also detect loss of voltage on one side and would close when its own pre-set time was reached, re-establishing supply to part of line B by back-feeding from line A. This is a simplified version of an actual system, which was a little more complicated in reality. The advantage of the closed loop system is that it does not rely on radio communications and is relatively fast in reconnecting customers. However, the system still relies upon customers who lose their supply notifying the power authority that there has been a fault on the line. The overhead line powered autoreclosers, which were used in this application could only accept power from one side, and had to be modified to include a low-voltage solenoid fed from external voltage transformers on each side, which made the scheme relatively very expensive.

As the operation of autoreclosers with magnetic actuators is independent from the overhead line, units of this type are ideal for closed loop application and any other application where system reconfiguration may require the direction of power supply to be reversed. Several manufacturers, including Cooper Power, The Joslyn Hi-Voltage Corporation, FKI and ALSTOM, offer autoreclosers powered by magnetic actuators. An example of the FKI unit is shown in Figure 11.13. This unit relies upon SF_6 gas for insulation, which has given excellent service but may be perceived to be a potential source of problems in the event of a gas leak from one of the gaskets that form part of its assembly. The issue of end-of-life disposal may also be of concern.

In order to avoid possible customer concerns over gas or oil leaks and end-of-life disposal, some manufacturers have introduced autoreclosers that use solid insulation and avoid the use of oil and SF_6 gas. Examples of these autoreclosers and their control units are shown in Figures 11.14 and 11.15.

The Joslyn TriMod™ Type 300 series autorecloser has ratings of up to 29.3 kV, 800 A and 16 kA. It houses a vacuum interrupter in each phase which has an aluminium outer body filled with a patented solid foam insulation. The unit uses an otherwise conventional vertical porcelain insulator and a removable polymer horizontal insulator. This latter insulator facilitates access for the removal of the phase current transformer.

The Kyle® Type Nova™ autorecloser is rated for use up to 38 kV, 800 A, 12.5 kA and consists of three solid polymer insulated interrupter modules, an encapsulated current transformer and an aluminium mechanism housing. It utilises state-of-the-art vacuum interrupters encapsulated in environmentally inert cycloaliphatic epoxy resin mouldings. The Nova cycloaliphatic epoxy is an outdoor polymer mixed with hardener and filler. This proprietary formulation is said to be superior in areas of surface tracking, hydrophobicity, ultraviolet light and physical properties.

The Kyle® Form 6 microprocessor-based relay (see Figure 11.15(a)) provides protective functionality and simple interactive graphical interfaces. Metering functions

Figure 11.13 *The FKI-type GVR autorecloser powered by a magnetic actuator (photo: author)*

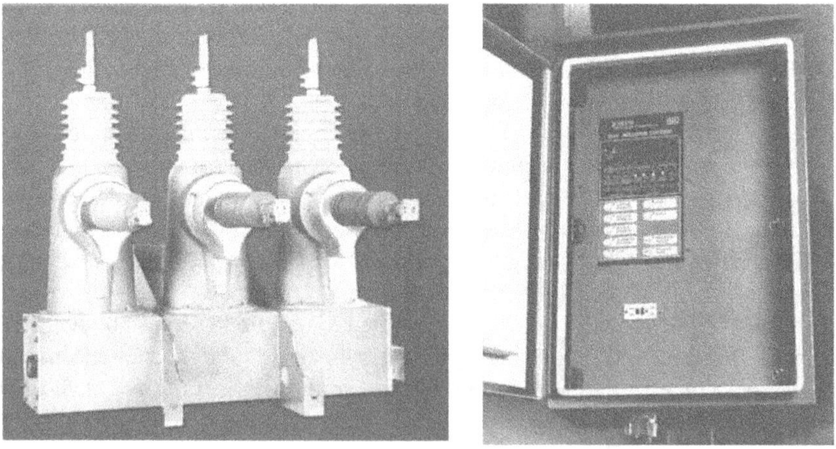

Figure 11.14 *The Trimod™ type 300 Series autorecloser and its control relay (courtesy of the Joslyn Hi-Voltage Corporation)*

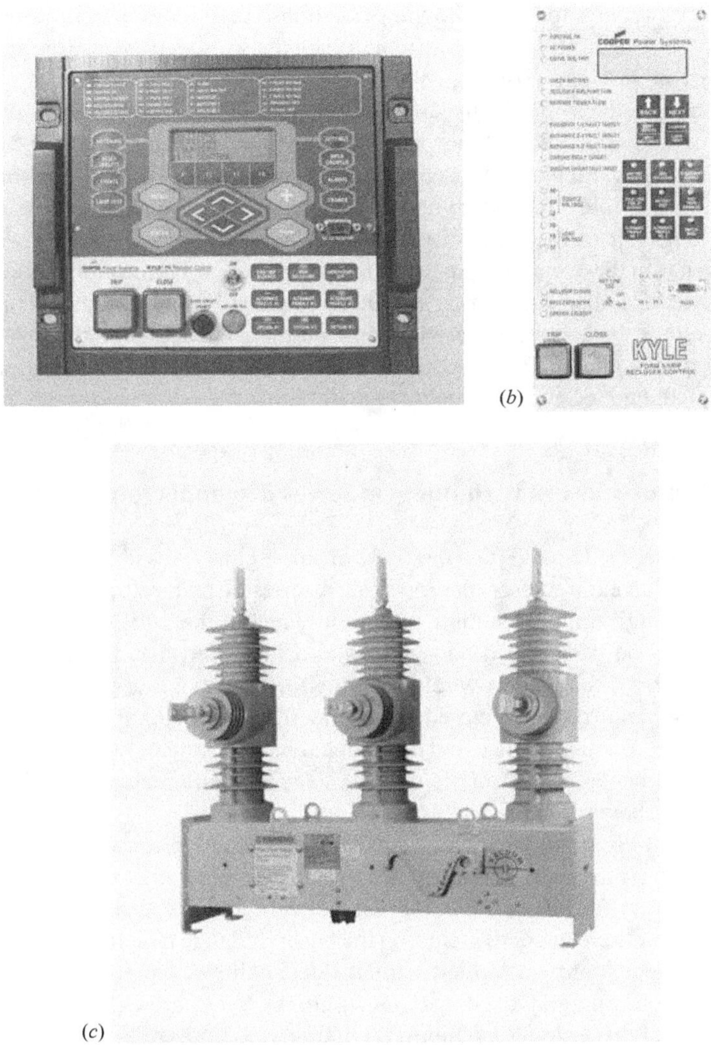

(*a*) (*b*) (*c*)

Figure 11.15 *The Kyle® type Nova™ autorecloser and control relays (courtesy of Cooper Power Systems)*

include demand and instantaneous current for each phase. Symmetrical components for both voltage and current are displayed along with kilowatt-hours for energy metering, second and 15th harmonics monitoring are also included.

The Kyle® Form 5 microprocessor-based autorecloser control (see Figure 11.15(b)), provides the intelligence to supervise an attached autorecloser. The control is also equipped with serial ports to interface with communication equipment. The control provides instantaneous and demand metering with programmable integration intervals for the following functions: real and apparent power in each phase, power factor, voltage and current in each phase as well as positive and negative sequence voltages, and positive, negative and zero sequence currents.

As was stated earlier, the use of microprocessors has meant that, in many ways, only the imagination sets a limit on the facilities that can be provided. Switchgear manufacturers, particularly those offering autoreclosers with solid insulation, are aware of the possible trade-off involved in using this insulation to avoid the use of oil or SF_6 gas. It is realised that solid insulation does not have the self-healing properties of oil or SF_6 gas in the event of voltage induced breakdown. The design, therefore, has to prevent electrical discharge on, or within, the insulation when subjected to excessive system voltage excursions, otherwise some permanent damage to the insulation will occur. If this is allowed to happen, further system voltage excursions may lead to accumulated damage and eventual insulation failure.

11.15 Autoreclosers with integral series disconnectors

Recently, a market demand for a new type of autorecloser was identified. The requirement was for an autorecloser incorporating a series disconnector within its assembly to provide a high-impulse voltage withstand when in the OPEN position. This was considered to be desirable because users had identified that an autorecloser with a built-in series disconnector would save them from having to purchase, install and maintain a separate open structured air-insulated disconnector. It also guaranteed safe operation because of electrical and mechanical interlocking between the autorecloser and its disconnector. An example of one of the first autoreclosers of this type, the type VPR, is shown in Figure 11.16.

The type VPR autorecloser uses specially developed vacuum interrupters, which are driven by a magnetic actuator within a stainless steel enclosure. While the unit contains SF_6 gas for insulation purposes, there are no gaskets involved in sealing in the gas. All joints are welded, including the bushing flange tank interface. In addition, drives into the unit are via welded stainless steel bellows.

The drive for integral disconnectors within the VPR autorecloser is manual only, via an external drive shaft (see Figure 11.17).

As the disconnector is an off-load switching device, the design of the contact drive mechanism is very important in terms of the contact make and break sequence of operation in relation to that of the main contacts. It has to be arranged that during the closing operation of the disconnector, its contacts are driven to the fully home position before the vacuum interrupter contacts close to complete the electrical circuit. This must, of course, allow for any pre-arcing. Similarly, during an opening operation, the vacuum interrupter contacts must open and clear the circuit before the disconnector contacts start to separate. This is achieved within the type VPR autorecloser by a cam drive to the disconnector, which is in effect a lost motion drive device.

The alternative VPR control relays are shown in Figure 11.18.

The two types of control relay are used with the VPR autorecloser are shown above. The relay shown in Figure 11.18(a) was designed to include the necessary

Figure 11.16 *A type VPR autorecloser with integral disconnectors and surge arrestors during installation (courtesy of ALSTOM T&D Ltd)*

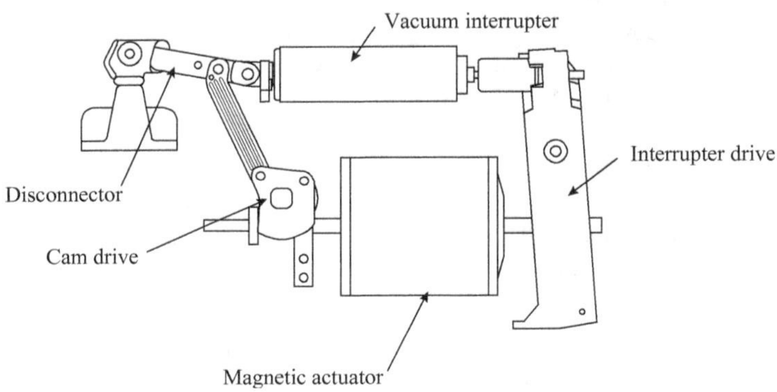

Figure 11.17 *Arrangement of internal disconnector drive of the type VPR autorecloser (courtesy of ALSTOM T&D Ltd)*

interfaces for remote control and monitoring, while Figure 11.18(b) is intended for local control applications. Both relays offer similar features, such as fault protection, current measuring and fault history, as described earlier for similar units.

11.16 A summary of the development of autoreclosers

Table 11.1 gives a summary of the chronological development of autoreclosers in terms of the order in which features were introduced.

(*a*) (*b*)

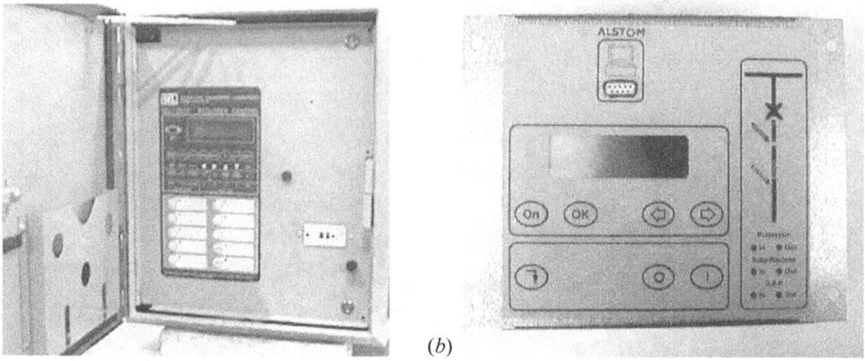

Figure 11.18 *The alternative microprocessor-based control relays VPR unit (courtesy of ALSTOM T&D Ltd)*

Table 11.1 *Chronological development of autoreclosers*

Operating power	Potential energy of external weight	✓						
	High voltage solenoid		✓	✓	✓			
	Magnetic actuator and lithium batteries					✓	✓	✓
	Magnetic actuator and rechargeable lead acid batteries						✓	✓
Primary internal insulation	Oil	✓	✓		✓			
	SF_6			✓		✓		✓
	Solid						✓	
Arc control	Arc control pot	✓	✓					
	Rotating arc SF_6			✓				
	Vacuum				✓	✓	✓	✓
Protection	Hydraulic	✓	✓		✓			
	Micro-electronic			✓	✓	✓	✓	✓
Series disconnector								✓

time⟶

11.17 Significant trends

There are four recent trends that are believed to be very significant in the field of autoreclosers. These are as follows:

(1) The application of remote control and automatic operation to pole mounted equipment is expected to increase, in order to maximise the continuity of supply by real-time reconfiguration of the network.

(2) Some manufacturers believe that there will be a customer preference for non-oil, non-SF_6 autoreclosers. If this belief proves to be the case, then units of this type will capture a greater market share. However, this share may be influenced by the introduction of designs, such as the stainless steel all welded SF_6 autoreclosers, which address the concerns regarding possible insulation leaks.

(3) Many users have identified that an autorecloser with built-in series disconnector will save them from having to purchase, install and maintain a separate open structured air-insulated disconnector. The first autorecloser of this type was produced by ALSTOM and the current model is known as the type VPR. There may be a growing customer preference for composite units of this type.

(4) Autoreclosers owe their existence to the need to detect and clear transient faults, and these transient faults can be avoided by using overhead cable instead of bare overhead conductors, which obviated the need for autoreclosers. Overhead cables have been used in Japan for at least the last 20 years and in Scandinavia for the last decade. At least two Regional Electricity Companies, RECs, in the UK have installed elements of overhead cable in their networks. It is the policy of one REC to use overhead cable for the main radial feeder, and in order to reduce costs, only use bare overhead conductors on spur lines. It is at these connecting points that they are installing autoreclosers with integral series disconnectors, to give a point of isolation. If this trend grows, the effect will be to reduce the normal and short-circuit current ratings of autoreclosers that will be required. This is because the installation point for autoreclosers will only be at spurs from the main radial feeder within the network, at the intersection of overhead cables and bare conductors, and will usually be some distance from the line incomer, which means that the impedance of the overhead cable will significantly reduce the fault level.

Chapter 12
High-voltage fuse-links

High-voltage current limiting fuse-links are widely used for the protection of distribution cables and transformers. They commonly form part of fuse-switch combination units, or in some cases they are used as stand-alone devices to provide the sole protection of equipment. Their particular advantages are their low first cost, their small dimensions and their ability to limit the peak fault current and let-through energy of a short-circuit fault to a small fraction of the prospective value. Well-designed fuse-links can limit the fault energy to around one 500th of what a conventional circuit breaker would let through. High rupturing capacity (HRC) fuse-links, therefore, give the applications engineer the opportunity to limit the damage which would otherwise result in the event of a short-circuit fault. The disadvantages of the HRC fuse-link include the necessity of stocking and carrying spare fuses and the need for manual intervention to replace a fuse, or fuses, in the event of a fault. In addition, unlike a circuit breaker, they are unable to detect zero sequence currents and, therefore, they will not operate on an earth fault that has a magnitude which is less than the rating of the controlling fuse-link. However, many of the applications for controlling transformers and cables in secondary electrical distribution circuits successfully employ HRC fuse-links.

12.1 Construction

A cross-sectional diagram of a typical high-voltage HRC fuse-link is shown in Figure 12.1. The fuse body is made from a strong, high-grade ceramic material, capable of withstanding the internal heat, thermal shock and pressure which would develop during a short-circuit clearance operation. The end caps and terminals are usually of plated brass or copper and are used to carry electrical current into and out of the fuse.

The fuse elements are specially designed wires or strips of silver, or copper, which serve to carry the normal rated load current of the fuse without deterioration, but will melt and interrupt the circuit in the event of the current rising to a pre-determined

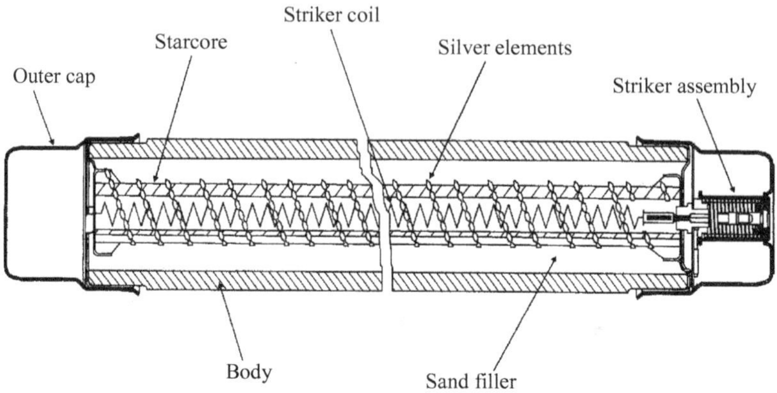

Figure 12.1 Cross-section of a typical high-voltage HRC fuse-link (courtesy Cooper Bussmann, Cooper (UK) Ltd)

level. The length of these elements has to be proportional to the system voltage in which the fuse is intended to work. Typically, this will be about 6 cm per kV. In order to avoid making the overall length of the fuse excessive, the elements of the fuse are usually coiled helically onto an insulated ceramic former within the fuse body. The wires or strips which form elements are designed to have a series of notches or holes along their length, and these, together with the cross-sectional area and number of elements connected in parallel, determine the electrical characteristics of the fuse for a given application.

Special measures are taken to ensure that the fuse elements do not need to reach their melting temperature of 1000°C in order to interrupt when operating on a low-level fault current. This is usually achieved by attaching a small quantity of a low melting temperature alloy, such as tin, or tin/lead/silver to each element notch to form an eutectic alloy that will melt at about 230°C. Thus, under almost any fault condition, the fuse will operate without excessive temperatures being reached, avoiding an adverse affect on its immediate surroundings. The effect brought about by adding an alloy to the fuse elements is called the 'M' or 'Metcalf' effect and is a feature of all high-voltage fuse-links manufactured in the UK, but not necessarily those manufactured elsewhere.

A very important part of an HRC fuse assembly is the granulated quartz-filling material. This is highly purified sand of closely controlled grain size. During normal service conditions, the filler material has the effect of conducting heat away from the fuse elements, to the walls of the fuse body and hence to the surroundings by convection. During fault interruption, the filler has an even more important role in extinguishing internal arcing and providing a current limiting feature to circuit interruption. See the later section dealing with fuse operation in service.

One further important feature of a high-voltage current limiting fuse is the striker pin assembly. This is a small internal unit attached centrally, within the fuse diameter, to one end of the fuse. The striker pin assembly is connected electrically to the opposite fuse end cap by means of a high-resistance wire, usually nichrome. In the event of a fault, the main fuse elements will melt and the current will be shunted into the resistance wire. This will either ignite a chemical charge within the striker pin, causing it to be propelled forward, or release a powerful spring, which will, in turn, propel the striker pin forward. In either event, the effect will be to cause a metal pin to be expelled from the end of the fuse-link. This can act as a simple visual indicator of fuse operation, or more commonly, operate a tripping mechanism to open all three phases of a switch. The importance of this is explained in the next section which deals with operation in service.

12.2 Operation in service

High-voltage HRC fuse-links are designed to carry their rated current for an indefinite period without deterioration, provided that there are no excessive overload currents. There are many well documented cases of fuses being examined after 25–30 years of load carrying service and showing no sign of deterioration. It can, therefore, be expected that, under normal service conditions, fuses should have a service life equal to or greater than the associated equipment.

It is important to note that the normal current rating marked on an air-insulated fuse-link is its rating in free air. Similarly, oil-insulated fuse-link ratings are based on oil immersion conditions. The fuse rating in service will be influenced by its immediate environment, in terms of the proximity of fuses in the other phases of three-phase units, and the fuse mounting attitude. These factors will influence heat dissipation and will reduce the normal current rating of the fuse-link. The manufacturer of the associated switchgear equipment will declare the fuse normal current rating to be applied. This would normally have been determined by carrying out temperature rise tests on the fuse-switch combination.

During the operation, to clear a major short-circuit fault, the temperature of the fuse elements will be brought up to melting point within a few milliseconds. This will cause the fuse elements to start arcing at the many points along their length where they have a reduced cross-section. These series arcs will be quickly quenched by the surrounding silica sand filler material, which is tightly compacted around the fuse elements. The silica sand filler will solidify into an amalgam of glass and partly fused sand, known as fulgurite, which has excellent insulating properties. The sudden rapid rise in circuit impedance results in the fault current being driven down towards zero well within the first loop of fault current. This sudden collapse of fault current flowing through the inductance in the circuit will produce a voltage spike. Fuse standards require that the fuse is designed to limit this voltage to a maximum specified value, which for a 12 kV fuse is 38 kV peak.

The current limiting effect of HRC fuses makes them virtually independent of the system fault level. The fuse will never experience the peak of a short-circuit current because it will have interrupted the circuit long before the peak current is reached (see Figure 12.2).

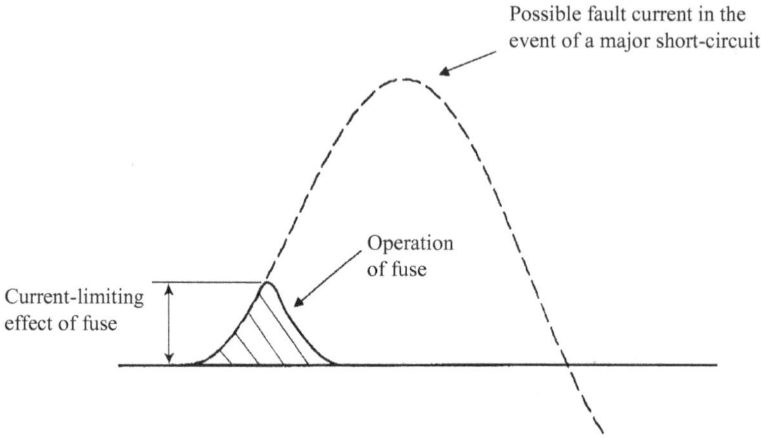

Figure 12.2 Cut-off of an HRC fuse on fault current operation (courtesy Cooper Bussmann, Cooper (UK) Ltd)

The higher the system voltage applied to the fuse, the more difficult are the internal arcs to extinguish and the longer the formation of fulgurite that takes place, hence the relationship between element length and applied system voltage. The formation of fulgurite sets a fairly close limit on the voltage rating of an HRC fuse. If a fuse is used within a system having a higher voltage than its rating, there may well be a failure to interrupt. Although fuses should never be used at voltages greater than their rating, they can be used at lower voltages. The only limit set on the lowest voltage at which they can be used is the voltage spike generated by interruption, which may lead to failure of external insulation. For example, insulation failure may occur if a 12 kV fuse is used on a 3.3 kV system. A general rule is that it is safe to use HRC fuses on systems having a voltage down to about half of the fuse voltage rating.

Although HRC fuses have an excellent ability to interrupt fault currents of high value, they can experience difficulty in interrupting overcurrents of low value. Fault currents of about three times the normal current rating of the fuse can be at the level where some difficulty in clearing the fault may occur. This difficulty is because at such low levels of current the fuse element pre-melting time can be up to several seconds, or even minutes. This causes the fuse elements to heat up unequally so that instead of the elements melting simultaneously at all of their series notches, only one or two breaks may occur at the hottest spots. This results in a small rise in impedance which is insufficient to give arc extinction, so the fuse is unable to interrupt the fault current. The smallest value of fault current that a high-voltage fuse can safely interrupt is known as the 'minimum breaking current' and the value of this is usually supplied by the fuse-link manufacturer on request. Fuses that are subject to this low overcurrent limitation are classed as 'back-up fuses' or in the USA as 'partial range fuses'.

In practice, the problem of having a minimum breaking current is completely overcome by using the fuses as part of a fused switch unit. The fuses used in such applications are fitted with striker pins. During fuse operation, these pins are ejected and actuate the three-phase tripping mechanism of the associated switch, which causes series connected switch contacts to open, interrupting the circuit. This sequence of events takes about 0.1 s to complete, whereas the fuse itself, operating at less than its rated breaking current, would take 1 s or more to reach failure point. The three-phase tripping mechanism also prevents single-phasing and burnout taking place in the event of a single-phase fault.

12.3 Fuse characteristics

12.3.1 Time–current characteristics

High-voltage HRC fuses have inverse time–current characteristics similar to those for circuit breakers and relays. There are no standardised characteristics for high-voltage fuses and the curves will differ from one manufacturer to another. Typical time–current curves are shown in Figure 12.3.

The above curves are plotted on a log–log graph. The convention outside of the USA is for the curves to represent mean values with a tolerance of less than ±20 per cent, and the curves are drawn dotted for values of less than the minimum safe breaking current. In the USA, separate minimum pre-arcing and maximum clearing curves are used.

12.3.2 Cut-off characteristics

The cut-off characteristics of a range of fuses give the peak let-through current plotted against different values of prospective current. See the example given in Figure 12.4.

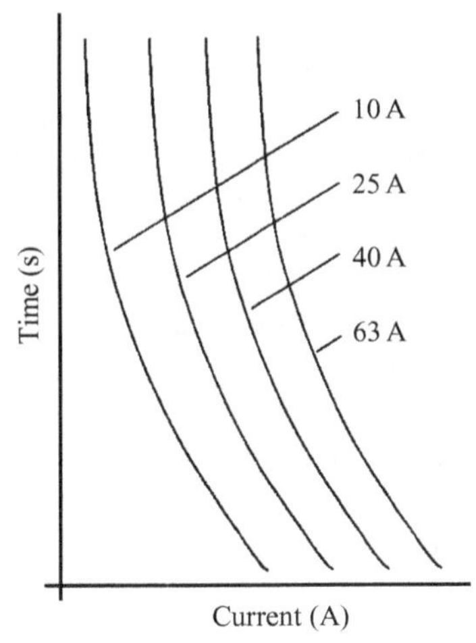

Figure 12.3 Typical time–current characteristics for a range of high-voltage fuses (courtesy Cooper Bussmann, Cooper (UK) Ltd)

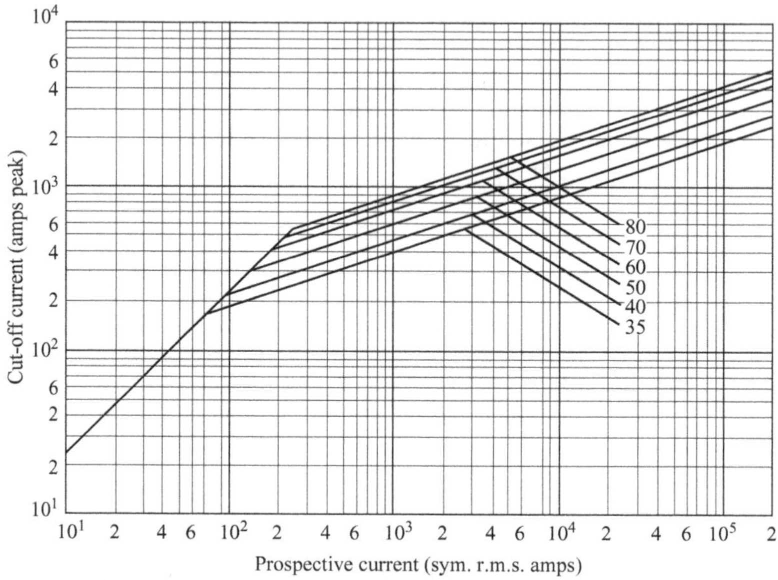

Figure 12.4 *Cut-off current characteristic of an high-voltage HRC fuse-link (courtesy Cooper Bussmann, Cooper (UK) Ltd)*

The cut-off current characteristic is used to ensure that a given fuse will adequately protect associated equipment when subjected to the highest envisaged fault current level.

12.3.3 The I^2t fuse-link characteristic

This is also known as the Joule integral and is a measure of the let-through energy that the fuse will allow during the process of fault current interruption. The units are ampere-squared seconds. Both pre-arcing and total clearing I^2t will be declared by the manufacturer, usually in graphical form. The pre-arcing I^2t is useful when checking protection discrimination between upstream and downstream fuses. The total I^2t of a downstream fuse-link must be less than the pre-arcing I^2t of the associated upstream fuse-link if damage to its elements is to be avoided.

12.4 Types of high-voltage HRC fuse-links

The photograph in Figure 12.5 shows three different fuse-link constructions.

12.4.1 British Standard oil-tight fuse-links

The British oil-tight fuses have standardised dimensions to BS2692-1. These fuses are fitted with special high-integrity oil seals to prevent the ingress of switch oil.

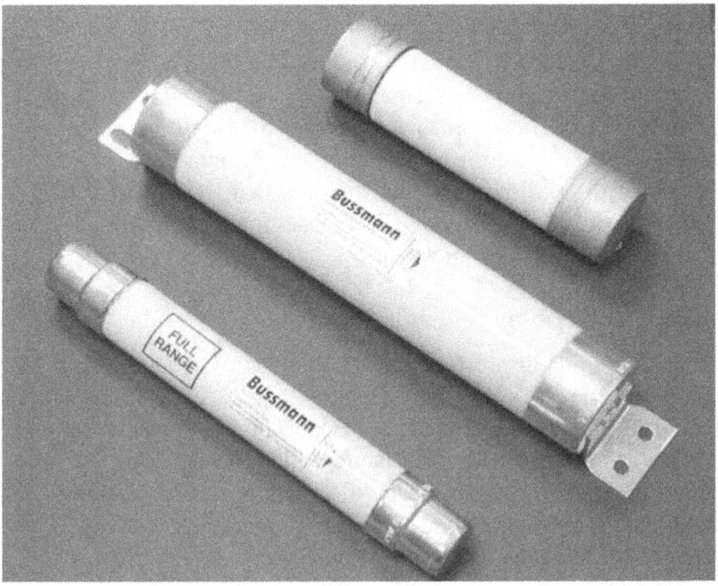

Figure 12.5 *Photographs of three different high-voltage HRC fuse-links (courtesy Cooper Bussmann, Cooper (UK) Ltd)*

Should such ingress occur, the fuse may rupture when trying to clear fault current. Until recently, this type of fuse has been, by far, the most widely used type in the UK and UK-oriented territories, mainly within oil-filled ring-main units. While the fuses have been designed for use in oil, they are not generally suitable for use in hot oil environments, such as transformer tanks. The current rating of these fuses is based on immersion in switch oil. Hence, if used in free air they require a substantial de-rating.

12.4.2 British Standard air-insulated HV fuses

British Standard air-insulated HV fuses come in a wide variety of types for both ferrule and bolted tag fixing. Ratings up to 72.5 kV are produced but, at the present time, at voltages above 11 kV they tend to be only used for more specialised applications.

12.4.3 DIN Standard air-insulated HV fuses

These fuses are manufactured to the DIN 43625 standard dimensions. They have a 45 mm ferrule at each end and the body lengths are standardised according to voltage rating. These fuses have a spring operated striker pin rather than the chemical type that are more usually used in British Standard practice. The DIN striker pin provides a tripping energy of 0.6 J, which accords better with continental switchgear designs. The British Standard striker pins provide a tripping energy of 2 J.

12.4.4 Motor circuit fuses

These fuses are specifically designed for the back-up protection of motors and their associated control equipment. The elements of a motor circuit fuse usually have stress relief bends along their length to enable them to withstand the alternate expansion and contraction they experience during the start up and running of direct-on-line motors. They are manufactured to the dimensions of the British Standard, DIN and North American practice. It is normal to have these fuses mounted within the motor control cubicle on the line side of the switchgear.

12.4.5 Instrument voltage transformer fuses

Instrument voltage transformer fuses are simple small diameter HRC fuse-links of low current rating, typically 3 A. These fuses are connected in series with instrument voltage transformers and are intended to provide isolation for the system in the event of a fault within the transformer. In practice, they have tended to introduce more problems than they solve, due to electrical discharge in the air space between the fuse and its housing in the cramped space that is normally available. They are not commonly fitted to continental switchgear unless specifically requested.

In addition, as the maximum normal current that can be carried by an instrument voltage transformer is of the order of 200 mA, and that under ferroresonance conditions is a current of about 400 mA, a 3 A fuse will not provide any protection against internal burnout.

12.5 Full range HV fuses

As distinct from the back-up or partial range fuses described earlier, these fuses embody more recent technology which extends the minimum safe breaking capacity down to the lowest value that can melt the fuse elements. Such fuses can, therefore, be used as sole protection in simple, low-cost housings, without the need for a striker pin tripping mechanism to handle low fault current levels.

A common and effective method of ensuring full range performance is to include a special 'miniaturised' version of an expulsion fuse in the fuse body assembly with the main current limiting elements (see Figure 12.6).

Within the full range fuse, the expulsion element handles low level fault currents up to five to ten times the normal current rating, and the main elements take over to clear higher fault currents as in a conventional back-up fuse.

12.6 Fuse standards

IEC 60282-1, which equates to BS 2692-1, is the general standard for all high-voltage current limiting fuses. It is a large and comprehensive document and covers definitions, performance, test parameters and gives a fuse application guide.

IEC 60787 is a useful guide to the selection of high-voltage HRC fuses-links for transformer protection applications. The standard sets out the various factors that need

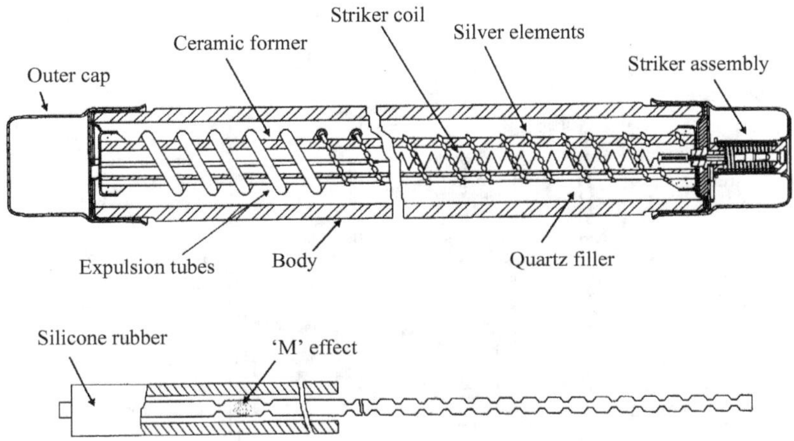

Figure 12.6 Cross-section through a typical full range high-voltage fuse-link (courtesy Cooper Bussmann, Cooper (UK) Ltd)

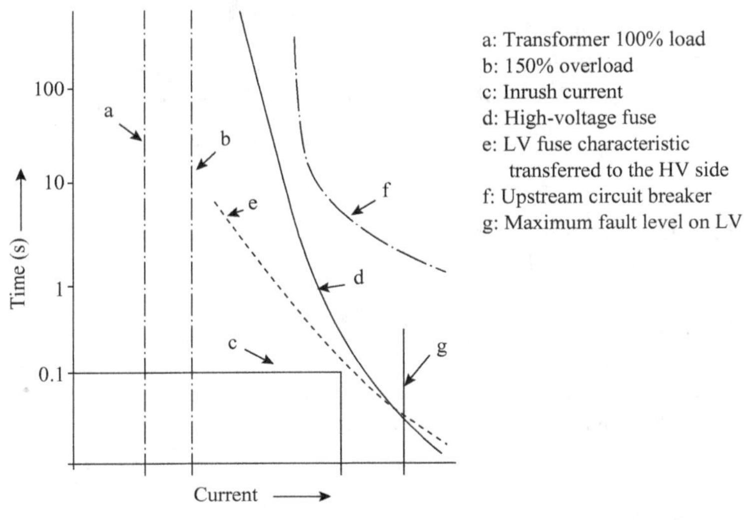

Figure 12.7 Discrimination of circuit breaker, HV fuses and LV fuses

to be considered when choosing an HV fuse-link for a given application involving an upstream circuit breaker and transformer low voltage fuses (see Figure 12.7). Note: The transformer inrush current is taken as being equal to 12 times the full load current for 0.1 s.

IEC 60644 relates to HRC fuses for the protection of HV motor circuits. The standard details the special withstand tests with which motor protection fuses need to comply and gives advice on the selection of such fuses for given applications. A fused motor control unit for direct-on-line motor starting is shown in Figure 12.8.

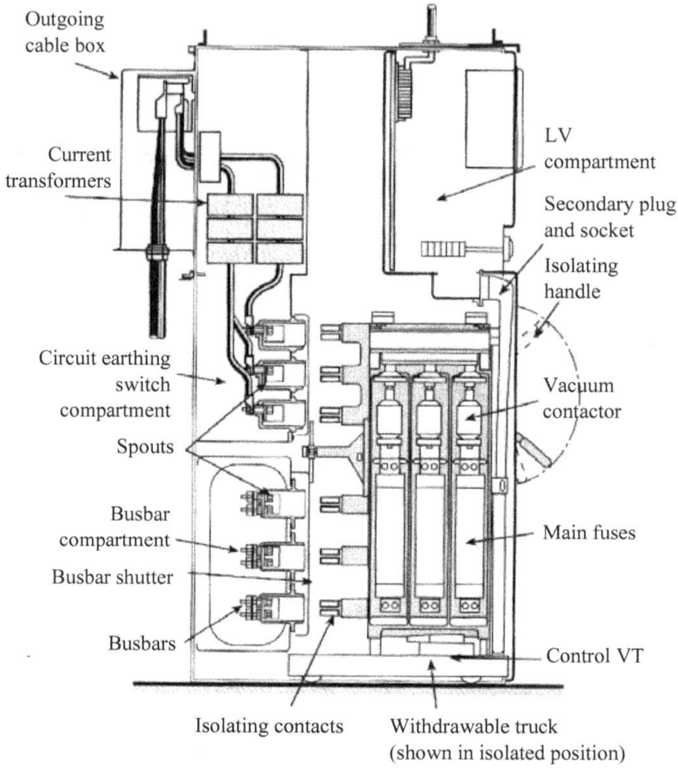

Figure 12.8 Cross-section through a type HMC1172 direct-on-line starter (courtesy of ALSTOM T&D Ltd)

IEC 605449 is the standard for high-voltage fuses that are to be used for the protection of capacitors.

IEC60282-2 is the general standard for non-current limiting high-voltage fuses. This standard also covers outdoor types of expulsion fuses, which are widely used.

12.7 Distribution applications

A large number of high-voltage HRC fuses are used for distribution system protection within secondary substations. They are to be found in a great variety of fused switchgear types, an example of which is shown in Figure 12.9.

IEC 60470 covers the various tests and test exemptions which apply to the use of particular fuse types and ratings within any given fuse-switch unit. The main areas of concern are:

(1) The I^2t and cut-off current values of a given fuse must not be greater than those which the fuse-switch has been proven by test to be capable of withstanding.

Figure 12.9 *A ring-main unit, type T4GF3, fitted with high-voltage HRC fuses (courtesy of ALSTOM T&D Ltd)*

(2) The transfer current, which is the maximum current that the switch has to clear unaided by the fuse during striker tripped operation, must be within the capability of the switch to clear without incurring damage.

(3) The thermal performance of the fuse within the environment of a given fuse-switch must be acceptable. Appendix F of IEC60282-1 gives the methods to be used in order to determine the de-rating that will be required to avoid excessive temperature rise and element deterioration in service.

12.8 Future trends

The 1980s saw the introduction of the first SF_6 ring-main units. Most manufacturers decided that, as it was not practicable to house high-voltage fuses within the gas enclosure because of fuse replacement problems, they would dispense with the fuses and produce units having a circuit breaker in the T-off section to control and protect the transformer. Other manufacturers decided to retain the high-voltage fuses, but house them in specially sealed enclosures to protect them from the environment. This situation still remains today and the use of T-off circuit breakers in ring-main units has reduced the demand for high-voltage fuses.

Some electricity distributers have introduced simple three-phase box housings using full range fuses without an associated striker pin operated switch. The effect of losing a faulty phase due to fuse operation, leaving the other two phases energised, may prove to be expensive if equipment burn out due to single phasing is the result. Time and experience will prove whether this solution has long term viability.

Chapter 13
Switchgear type tests

It is not the purpose, or possible, within the scope of this book to provide a detailed step-by-step guide on the procedures to be followed when conducting development and type testing of switchgear. Indeed, such a guide would run into several volumes and, due to detailed changes in specifications that take place from time to time, it would quickly become obsolete. It would also have to cover all types of switchgear including circuit breakers, switches, fuse switches, fuses, earth switches and disconnectors. However, the intention here is to give an overview of the important stages in testing that have to be completed in order to provide switchgear that is safe and able to operate correctly in the circuits and ambient conditions for which it is intended. By way of example, particular reference will be made to circuit breakers. The opportunity will be taken to highlight certain aspects of each type test in order to show the relative importance and the way in which solutions to encountered problems can, and usually will, have an influence on the performance of other type tests.

13.1 Reports and certificates

Testing laboratories issue two forms of test document which detail the tests that are conducted, giving the test parameters and the equipment performance. The documents have specific names to avoid confusion. The first is a 'Report of Performance' and the second is a 'Certificate of Rating'. There is a significant difference between these documents that must be understood, as superficially they are similar in appearance. It is clearly important that those who are not familiar with them understand the difference when evaluating the suitability of switchgear for a specific application. The similarity in appearance of both types of document can be seen in Figure 13.1.

Figure 13.1 shows the front sheet of a Report of Performance on the left-hand side and a Certificate of Rating on the right. Both documents have a hard-back binding and have the testing station seal, but the title wording identifies each type of document. The appearance of each is substantially similar, demonstrating how the inexperienced examiner could be confused.

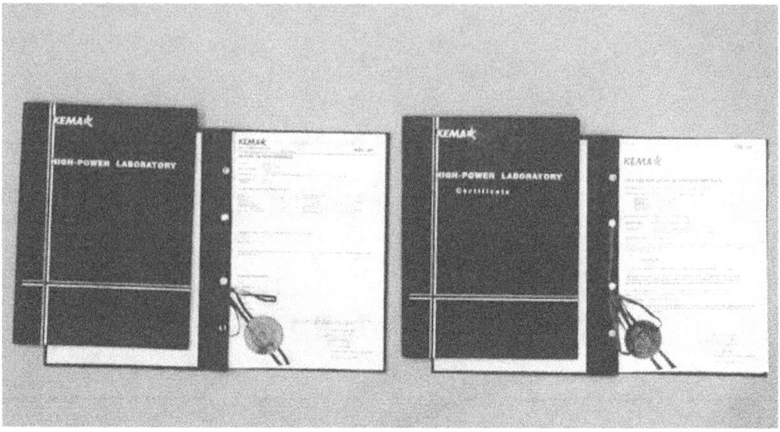

Figure 13.1 *Report of Performance and Certificate of Rating documents (photograph courtesy of the KEMA Laboratories)*

The conditions under which the test authority will issue these documents is explained in the following:

The testing authority will issue a 'Report of Performance' when

(1) there is no specific national or international standard covering the equipment tested;
(2) the values used for the tests were not in accordance with a specific standard;
(3) the complete tests in the specification were not carried out to the letter;
(4) the equipment fails some aspect of the tests in the standard.

The testing authority makes it clear that when copying a 'Report of Performance', the whole of the document must be copied, not just the front sheets. It should be noted that the report might not include any or full identification of the equipment tested, and the testing authority will not necessarily verify any identification that is included.

A 'Certificate of Rating' is issued by the testing authority when the equipment tested meets all of the requirements of the standard, and the associated STL guide, in every detail. It has been known for a 'Certificate of Rating' to be withheld and a 'Report of Performance' to be issued in its place because of an anomaly in the post-test no-load travel records. Only the front sheets of a certificate need be copied for performance evidence, as the testing authority puts its reputation behind the validity of the claimed performance. The complete 'Certificate of Rating' must contain sufficient information to accurately identify the equipment tested, which the testing authority verifies at the time of test.

A 'Report of Performance' should not, however, be dismissed. The question that the reviewer needs to ask is 'why was a certificate not issued?'. If a satisfactory explanation is given, the reviewer will then need to examine, in detail, every aspect of the report to ensure that all aspects, such as transient voltage recovery, current, peak current, d.c. component and time intervals in a sequence, meet the specified values.

13.2 National and International Standards

Most countries have produced their own National Standards for switchgear. For example, the British Standards Institute has a standard covering a.c. circuit breakers. In addition, some 22 countries have taken an active part in producing an International Standard covering the same equipment and voltage range, under the auspices of the International Electrotechnical Commission, known as the IEC, which is based in Geneva. In addition, end-user standards such as the Electricity Association Technical Standard EATS41-36, 'Distribution switchgear for service up to 36 kV (cable & overhead conductor connected), issue 1, 2000', which defines the specific requirements of the UK, add to the International Electrotechnical Commission standards, those requirements which are seen as being necessary in the UK. The complete list of reference standards relating to switchgear is formidable but the most important are listed in Chapter 18.

Other switching equipment, such as contactors, earth switches, disconnectors and fuse-switch combinations, have their own specific standards. These also share some related standards with circuit breakers. These shared requirements are published in an IEC 'Common clauses' document.

It should be noted that all standards, including those produced by the IEC and Electricity Association are all under consideration for modification at any time. It is, therefore, very important to ensure that the latest edition is used.

13.3 Development tests

Development tests are carried out by manufacturers to allow the performance of design concepts and completed prototypes to be explored. These tests fall into four categories:

(1) test rig work;
(2) exploration of the limits of performance;
(3) proving a new product before certification, or proving changes to an established product before re-certification;
(4) certification tests.

Taking each in turn:

(1) At the outset of the development of a new product, the design team will identify those areas of the design which may contain a measure of technical speculation. Test rigs will then be designed and built so that tests can be carried out in order to be certain that working solutions have been found and that these solutions can be built into the prototype design. These test rigs can be produced to look at any area of the design, thermal, mechanical or electrical, and will usually include measuring transducers which, of course, do not form part of the eventual product that will be type tested.
(2) While some test rig and development testing will be carried out at in-house test facilities, development tests to explore the limits of performance may need to

be conducted at independent testing stations such as KEMA in the Netherlands, CESI in Italy and BSTS in the UK. As test rig and development tests are controlled by the manufacturer, and would not necessarily comply with the requirements of a specific standard, these testing stations would issue Reports of Performance detailing the test results obtained whether or not the results were successful.

(3) During the course of test rig and development tests, changes may be made to the design in order to optimise the performance. When individual aspects of performance have been proved in isolation, it is very important to prove the complete design, before certification. This is to check for any unforeseen interaction between the previously unlinked components. Similarly, the influence of design changes to an established product should be established by test if problems in service are to be avoided.

(4) Certification tests will be carried out by an approved independent testing authority. The term 'independent' means that, while it may be owned by a manufacturer, the testing station will have complete technical independence and the test results will remain confidential between the testing station and the client, usually a manufacturer, paying for the tests. The testing station will control all aspects of the tests to be conducted without any interference from the client, although it is normal for them to attend and witness the tests. These testing stations will issue a Certificate of Rating to a nominated standard, however, for this to be issued all aspects of the tests must be faultless, and to the letter of the standard.

Complete type tests on a circuit breaker will include the following tests:

(1) Mechanical operation.
(2) Temperature rise.
(3) High-voltage
    power frequency, dry, on the primary circuit;
    power frequency, wet, on exposed insulated designs;
    power frequency on secondary wiring;
    basic impulse level (BIL);
    partial discharge.
(4) Short-circuit
    terminal faults;
    low-inductive switching;
    capacitor switching;
    back-to-back capacitor switching;
    internal arc faults.
(5) Environmental tests
    temperature cycling;
    icing tests on outdoor switchgear.
(6) EMC
    electromagnetic compatibility.

We will look at them in more detail.

13.3.1 Mechanical operations

For conventional circuit breakers, this consists of a total of 2000 operating cycles carried out under defined conditions. For autoreclosers, this figure is 4000 operating cycles. The electrical integrity of the test circuit breaker is established at the beginning and end of the tests, and moving contact travel records are taken periodically during the test. Some of the operations are arranged to be of the make–break type; that is, the initiation of the opening trip operation is via the main contacts, which, for this purpose, are connected in the trip circuit.

At the end of the tests, the criteria to be met are:

(a) The operating travel characteristics at the end of the tests should be substantially the same as those recorded at the beginning.
(b) The electrical integrity of the test circuit breaker should be confirmed.
(c) A detailed examination of all parts should confirm that no undue distortion or wear has taken place.

The normal sequence of operations used for the test is given in Figure 13.2.

13.3.2 Temperature rise

It is a surprise to those who do not know, but nonetheless true, that the normal current rating of a circuit breaker is determined by the measurement of the temperature rise of critical components when passing the rated normal current. In this test, measurement is carried out by the attachment of thermocouples to the critical components and recording the results. For tests on a circuit breaker, if it is of the switchboard type, the test circuit breaker will have its external side sheets insulated with 25 mm thick

Operating sequence	Control voltage	Number of operating sequences	
		Circuit breakers for autoreclose duty	Circuit breakers not for autoreclose duty
	Minimum	500	500
$C-t_a-O-t_a$	Rated	500	500
	Maximum	500	500
$C-CO-t_a-CO$	Rated	500	–
$CO-t_a$	Rated	–	500

Figure 13.2 Table of rated sequences for the mechanical operations test

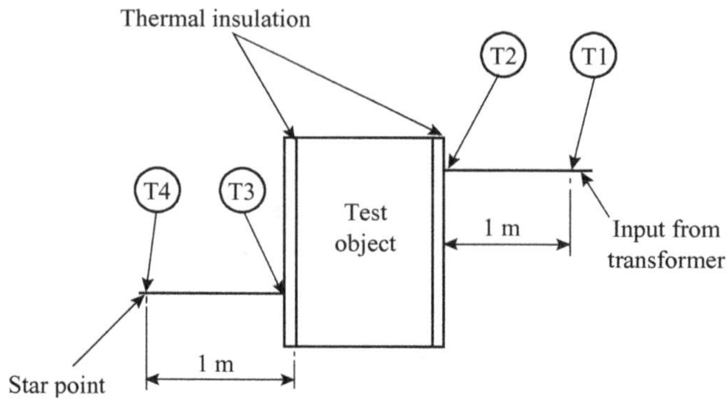

Figure 13.3 Temperature rise test arrangement

Type of connection	Maximum temperature rise (°C)
Plain copper contacts	55
Plain copper, bolted	65
Silver plated contacts	65
Silver plated, bolted	75

Figure 13.4 Table of maximum allowable temperature rise

polystyrene foam sheets. This is to ensure that no heat is lost to adjacent switchboard panels. The current used for the test is usually supplied at low voltage by a transformer and its value is also monitored. The connections to the test circuit breaker are by copper bars at least 1 m long, as are the connections to the star-bar. The temperature difference between the direct connection to the test circuit breaker and that measured 1 m away must be maintained within 5° to ensure that the test circuit breaker is not exporting heat. The ambient air temperature is also monitored (see Figure 13.3).

The maximum temperature rise recorded must be equal to, or less than, that allowed in the specification. These specified values depend on the type of connection, the materials involved and whether or not silver plating has been used. Refer to the relevant standard for a comprehensive list of allowable temperature rises, however, some examples of the specified maximum allowable temperature rises are given in Figure 13.4.

If the maximum temperature rise is in excess of that allowed, the switchgear designer will have to decide on the measures to be taken, and has two choices. He will either have to arrange to create less heat, although even a small excess in temperature rise will be very difficult to reduce by this method, or he will have to transfer the excess heat to either the atmosphere, with increased ventilation, or to other areas within the switchgear where the temperature is lower. If this latter method is used, care must be taken to safeguard compartment integrity in the event of an internal arc. Sometimes a matt black surface finish will be used to increase thermal emissivity of critical components. This will increase thermal emissivity by about only 2 per cent but this may be sufficient in some cases.

The temperature rise at each thermocouple is usually measured and recorded every hour on a data logger and the test is completed when a steady state is reached. This is defined as being when the measured temperature does not increase by more than 1°C in 1 h.

A switchboard type circuit breaker, having its sides clad with polystyrene foam sheet to prevent heat loss to adjacent panels, together with a data logger for continuous temperature monitoring and recording can also be seen as shown on test in Figure 13.5.

In the early stage of a development, there may be several arrangements that require testing and these tests may consume a great deal of time. When the test current is switched on, the temperature of components will start to increase from the starting ambient temperature. The temperature rise will follow a heating curve, which is mathematical in its nature. The maximum, or steady state, temperature of this heating curve can be determined early in a test by taking three temperature readings at equal time intervals and using these to calculate the maximum value that would be attained. A temperature rise curve with marked time intervals is shown in Figure 13.6.

Figure 13.5 A panel type circuit breaker on temperature rise test (courtesy of ALSTOM T&D Ltd)

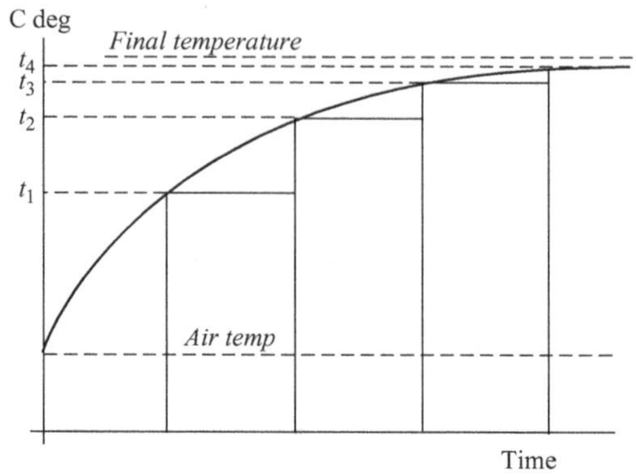

Figure 13.6 Temperature rise with equal time interval measurements

A heating or cooling curve is defined by the equation

$$T_\mathrm{r} = T_{\max}(1 - \varepsilon^{-t/\tau})$$

where T_r is the temperature rise, $T_{\max}$, the maximum temperature rise, ε, the Napierian log base, τ, the time constant and t, the time from start.

The heating curve in Figure 13.6 is mathematical, and the successive elements of time form a geometrical progression, where each term is equal to the preceding term multiplied by the common ratio. The total to infinity of such a series is

$$S = \frac{a}{1 - r}$$

where a is the first term and r, a common ratio.

Referring to Figure 13.6, if a circuit breaker is being evenly heated by the passage of electric current then it will gain a temperature of t_1 in the first interval and a further temperature t_2 in the second interval, which is equal to the first, then:

$$a = t_1$$

and

$$r = \frac{t_2}{t_1}$$

Thus

$$S = \frac{t_1}{1 - (t_2/t_1)}$$

$$= \frac{(t_1)^2}{t_1 - t_2}$$

(temperatures t_1 and t_2 are in °C).

More generally, if three thermocouple readings are taken at equal intervals of time, say t_1, t_2 and t_3, at any stage of a temperature rise test, the values obtained can be used to predict the maximum temperature rise that will be achieved.

$$T_r = \frac{t_2 - t_1}{1 - (t_3 - t_2)/(t_2 - t_1)}$$

Example 13.1

During a temperature rise test on a prototype circuit breaker, three thermocouple readings were taken at 20 min intervals. The readings were 21.5, 27.1 and 32°C. What is the predicted final temperature rise above the starting temperature?

Given

$$T_r = \frac{t_2 - t_1}{1 - (t_3 - t_2)/(t_2 - t_1)}$$

And substituting the thermocouple readings

$$T_r = \frac{27.1 - 21.5}{1 - (32 - 27.1)/(26.2 - 21)} = \frac{5.6}{1 - 0.875}$$

$$= \frac{5.6}{0.125}$$

Therefore, the temperature rise

$$= 44.8°\text{C}.$$

There is a relationship between the steady state temperature rises experienced by a circuit breaker for two different constant currents over the normal working ambient temperature range. This relationship is

$$\frac{T_1}{T_2} = \frac{[I_1]^\delta}{[I_2]^\delta}$$

The suffixes 1 or 2 denote the two test conditions.

This relationship is very useful for calculating the likely normal current that can be safely carried when a circuit breaker is to be installed in an ambient temperature in excess of the standard maximum of 35°C. As previously explained, the normal current rating of a circuit breaker is determined by the temperature rise above an ambient temperature when carrying a current equal to the rated value. This means that the actual maximum temperature allowed is the temperature rise added to the maximum allowed ambient temperature. If the circuit breaker is to be installed in a site where the ambient temperature is in excess of the maximum value, the allowed temperature rise must be reduced by 1°C for every degree by which the ambient exceeds the specified value. The following is a worked example showing how the new rating can be determined.

Example 13.2

An 11 kV circuit breaker was found on test to have a maximum temperature rise of 65°C when carrying a continuous current of 630 A. The same circuit breaker had a maximum temperature rise of 60.69°C when carrying a lower continuous current of 600 A. What would be the likely maximum permissible continuous current rating if the circuit breaker was to be installed on a site where the steady state ambient temperature would be 54°C?

Given that

$$\frac{T_1}{T_2} = \frac{[I_1]^\delta}{[I_2]^\delta}$$

Substituting the known test values into the expressions:

$$\frac{65}{60.69} = \frac{(630)^\delta}{(600)^\delta}$$

Therefore

$$1.071 = 1.05^\delta$$

Taking logs to the base 10 of both sides of the expression

$$0.0298 = \delta \times 0.0212$$

Therefore

$$\delta = \frac{0.0298}{0.0212} = 1.4$$

As the proposed ambient temperature is to be 54°C, the allowable temperature rise will be reduced by a value equal to 54°C minus the standard maximum ambient temperature of 35°C, giving a 19°C reduction, making the allowed temperature rise 65°C minus 19°C, equal to 46°C.

Substituting the known values into the given expression

$$\frac{65}{46} = \frac{(630)^{1.4}}{(I_2)^{1.4}}$$

Therefore

$$\sqrt[1.4]{1.4} = \frac{630}{I_2}$$

making

$$I_2 = 500\ \text{A}.$$

Therefore, the 630 A rated circuit breaker would have to be de-rated to 500 A if it had to operate in a steady state ambient temperature of 54°C.

Before leaving the subject of temperature rise tests, there are a number of important issues that are worth noting.

- It is usually more difficult and more expensive to produce less heat by increasing the cross-sectional areas than is initially thought. Even an excess temperature of only, say, 2°C or 3°C would be very difficult to eliminate.
- Within air-insulated switchgear, conduction and ventilation to cause heat to flow into cooler areas may offer a better solution.
- Important solid insulation in air-insulated switchgear should always be vertically aligned rather than horizontal to increase air flow and reduce the possibility of contamination which could lead to insulation failure.
- It is sometimes useful to use brass castings as part of the conducting circuit. Care should always be taken, both at the prototype and production stage, to ensure that the brass does not include any phosphorous, as even a minute quantity will drastically reduce its conductivity and cause an increased temperature rise.

13.3.3 High-voltage tests

Three types of test are used to determine the insulation integrity of switchgear. These are the power frequency, basic impulse and partial electrical discharge tests.

13.3.3.1 Power frequency tests

Power frequency tests, sometimes known as the high pot or flash tests, are carried out using a test voltage of about 2.5 times the rated voltage at the rated frequency. These tests are designed to stress the insulation of the test switchgear, both interphase, interpole and to earth. This test also is carried out on production units as a routine test. A wet power frequency test is carried out on switchgear intended for outdoor use with exposed insulation. For this test, a finely divided spray of water with a specified salinity is directed at the insulation while the test voltage is applied.

A 2 kV power frequency test of 1 min duration is carried out on the secondary wiring of switchgear. This test is also carried out on production units as a routine test.

13.3.3.2 Basic impulse tests

The basic impulse level (BIL) withstand test is designed to stress the insulation of switchgear by the repeated application of a very steep fronted voltage wave and is a very searching test of the unit's insulation integrity. The shape of the impulse voltage wave is specified to be 1.2 μs/50 μs, which means that it has to reach its specified BIL peak voltage value in 1.2 μs and have a wave tail which falls to half of the BIL peak value in 50 μs (see Figure 13.7).

The impulse voltage wave is provided by an impulse generator which is essentially a bank of capacitors which are charged up connected in parallel, and discharged in a series connection by the triggering of sphere gaps. A photograph of an impulse generator is shown in Figure 13.8.

The basic impulse test consists of 15 consecutive impulses, of both polarities, applied to each pole in turn with the circuit breaker open and all other poles earthed,

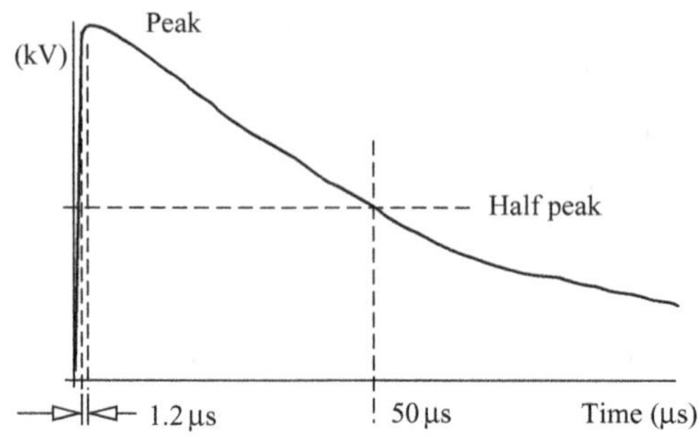

Figure 13.7 Standard basic impulse level (BIL) waveform

Figure 13.8 An impulse generator (photograph courtesy of the KEMA Laboratories)

and then to each phase in turn with the other phases earthed, as shown in Figures 13.9 and 13.10. The rules do not allow more than two disruptive discharges in any 15.

In order to ensure that all insulation, interphase, phase to earth and across contact gaps are proven, the eleven arrangements shown in Figure 13.10 must be tested.

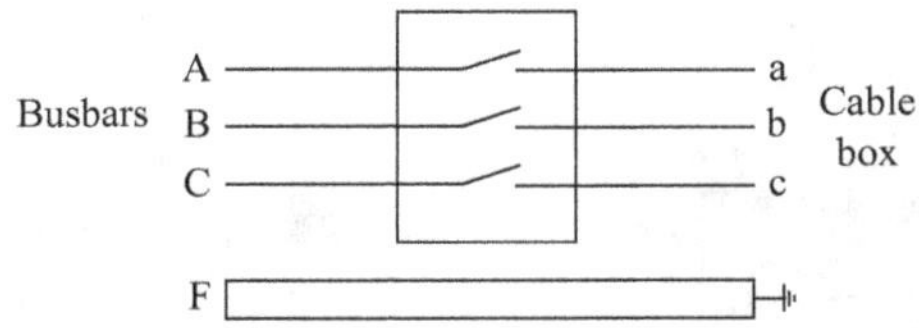

Figure 13.9 Terminal identification for high-voltage tests

Test sequence	Switching device	Voltage applied to	Earth connected to
1	Open	A	BC abc F
2	Open	B	AC abc F
3	Open	C	AB abc F
4	Open	a	bc ABC F
5	Open	b	ac ABC F
6	Open	c	ab ABC F
7	Closed	Aa	Bb Cc F
8	Closed	Bb	Aa Cc F
9	Closed	Cc	Ab Bb F
10	Not inserted	ABC	F
11	Not inserted	abc	F

Figure 13.10 Test arrangements

13.3.3.3 Partial discharge tests

Partial electrical discharge taking place within switchgear can, by the formation of nitric acid and nitric oxide, eventually lead to catastrophic failure. Electrical discharge can take place across stressed air gaps between insulation and materials at earth potential. Most manufacturers design their equipment to ensure that the levels of voltage at which discharge inception and extinction take place are well above the maximum value applied by the system voltage.

Discharge inception occurs on a rising voltage, and discharge extinction will take place as the voltage returns towards zero. However, inception always occurs at a higher voltage than extinction. In order to guard against electrical discharge continuing when the voltage returns to normal, it is usual to arrange for discharge extinction to be at least 10 per cent above that applied by the maximum system voltage. With these values, a modest system voltage excursion will not induce electrical discharge, and larger voltage excursions will automatically be extinguished once the system returns to normal. It will be seen that on an unearthed system, running with one phase down to earth, the inception and extinction voltages will need to be higher than those installed within an effectively earthed system, and the switchgear must meet these higher levels if discharge is to be avoided.

(*a*) Discharge free (*b*) With discharge

Figure 13.11 *(a) and (b) An ERA-type discharge detector (courtesy of ALSTOM T&D Ltd)*

Current specifications require tests to be carried out on components, but not on complete equipment. However, draft changes to National and International specifications are currently being discussed and it is likely that these tests will be included in the near future.

Discharge inception and extinction levels are measured during type tests by installing the switchgear in a Faraday cage and using a discharge detector (see Figure 13.11).

If no discharge is detected, the elliptical display on the measuring equipment will be undisturbed, as shown in Figure 13.11(a). When discharge is present, the elliptical display will show tearing and discontinuity as shown in Figure 13.11(b).

13.3.4 Short-circuit tests

Short-circuit fault testing is the most expensive form of type testing that has to be carried out. This is because the testing station is equivalent to an electrical power generating station, but with generators and test transformers designed to produce very high levels of short-circuit current necessary to reproduce fault conditions. Sophisticated measuring and adjustment of circuit power factor and current level is necessary, as well as point-on-wave fault initiation. Testing stations are therefore very expensive to construct and run. Some manufacturers have there own in-house short-circuit test facility, such as BSTS in the UK which are run as independent units having their own test observers, and other testing stations are independent of manufacturers, such as the KEMA Laboratories in the Netherlands and the CESI laboratories in Italy. The size and complexity of one of these testing stations can be seen in Figure 13.12.

13.3.4.1 Terminal short-circuit faults

National and International standards specify that three operations, in the circuit breaker rated sequence, shall be carried out at 10, 30, 60 and 100 per cent of the symmetrical fault current level using a specified rate of rise of recovery voltage at each level. Also at each level, the final two break operations should be preceded by a fault make operation. Indeed, the fault make operation is mandatory at the 100 per cent

Figure 13.12 Part of the KEMA test facilities in the Netherlands (photograph courtesy of the KEMA Laboratories)

symmetrical fault level. Figure 13.13 shows an oscillogram of test duty 4, 100 per cent fault rating test carried out as an O–CO sequence.

At first sight, the oscillogram shown in Figure 13.13 appears to be very complicated, but once the individual traces on the record are identified, the oscillogram becomes much easier to understand. It can be seen that the record gives a complete picture of all that happens during a Test Duty 4, O–CO sequence, at the 100 per cent rated fault level. The top two traces show the current in the trip and closing coil. Moving down the oscillogram, the voltages and current in each phase, R, Y and B, are shown. Superimposed upon these traces is the contact travel record. The trace at the very bottom of the record is designed to detect any transient earth fault current flowing to the circuit breaker enclosure.

The sequence of events, starting with the test circuit breaker contacts closed and no current flowing, can now be examined by starting at the left-hand edge of the oscillogram and moving to the right, in the direction of elapsed time.

- The first change we see is the appearance of fault current in each phase.
- Next, at the top of the record, current can be seen to flow through the trip coil. This current shows the characteristic dip that occurs when the coil armature moves, changing the coil impedance.
- Shortly after this, the moving contact record shows that the contacts have started to move towards the OPEN position.
- Part way through this contact movement the fault current can be seen to be extinguished and a recovery voltage appears across the contact gap in each phase.
- The contacts remain in the OPEN position, in this case for 300 ms, as the circuit breaker being tested is intended for use in autoreclose applications.

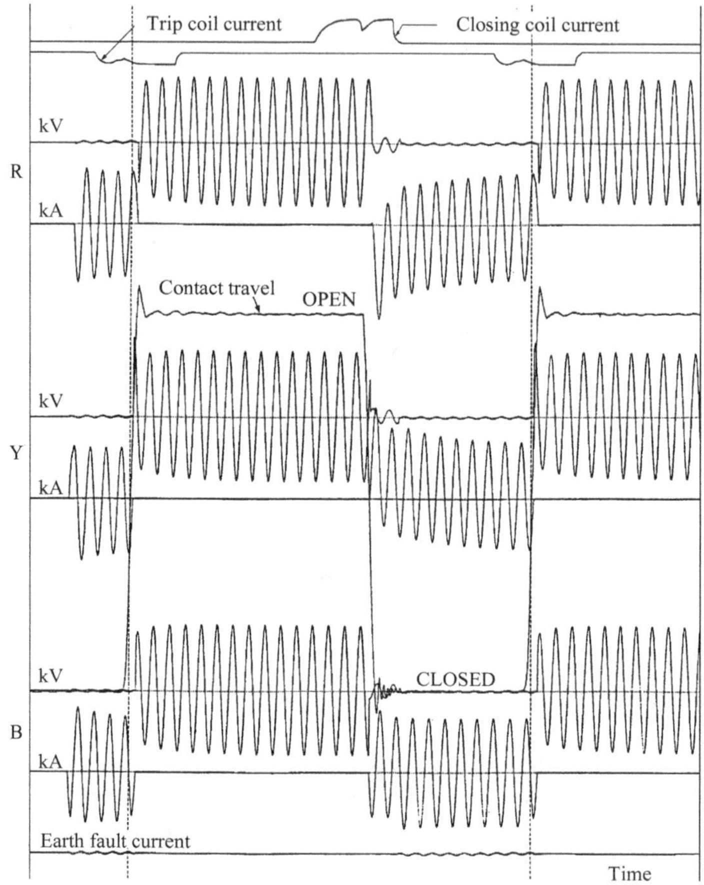

Figure 13.13 Oscillogram of Test Duty 4, 100% rated fault level (courtesy of the KEMA Laboratories, Netherlands)

- Towards the end of the 'dead-time', it will be seen that current starts to flow in the closing coil and the moving contacts move to the CLOSED position, causing fault current to flow.
- The testing station will energise the trip coil to initiate the trip operation at a time when the transient reactance of the fault current has decayed to not more than 20 per cent, which is the maximum specified value for this test duty.
- Once again the circuit breaker contacts will move towards the OPEN position, interrupting the fault current, and a recovery voltage then appears across the contact gap of each phase.

13.3.4.2 Asymmetrical breaking capacity, Test Duty 5

In addition to the symmetrical breaking capacity tests, which are Test Duties 1–4, tests have to be carried out with a level of current asymmetry having a d.c. component.

This d.c. component of current is a function of the opening time of the circuit breaker (see Chapter 4, Figure 4.4). The testing station will carry out these tests with the objective of trying to obtain a major extended loop in each phase in turn in order to cause as much distress to the circuit breaker as possible. An oscillogram of a Test Duty 5 interruption is shown in Figure 13.14 and having worked through the previous figure showing a symmetrical test, this test record should be easier to understand.

The sequence of events which are recorded are as follows:

- Starting with the test circuit breaker contacts closed, the testing station closes its point-on-wave fault making switches to initiate the flow of asymmetrical fault current. In this case, the current in the Red phase has maximum asymmetry.

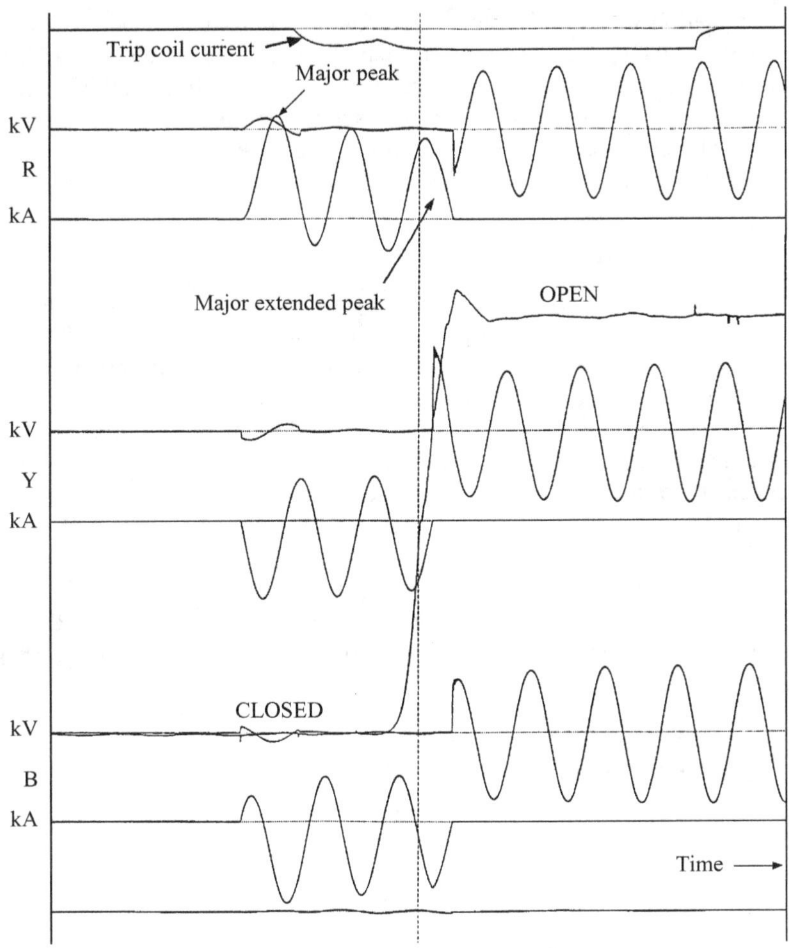

Figure 13.14 Oscillogram of Test Duty 5; asymmetrical fault level (courtesy of the KEMA Laboratories, Netherlands)

- The testing station then initiates current in the test circuit breaker trip coil to open the circuit breaker so that the rated degree of asymmetry will be present at the time of contact separation.
- The moving contact travel trace shows the start of movement towards the OPEN position.
- The contacts separate just before current zero in the Yellow phase, allowing that phase to clear with only a few milliseconds of arcing, causing a phase shift of current in the other two phases which make their values equal and opposite. Recovery voltage appears across the Yellow phase.
- The current in the Red phase, which has a high level of asymmetry, can be seen to be extended by the phase shift.
- The currents in both the Red and Blue phases then clear and a recovery voltage appears.

The testing station would then attempt to move the point of contact separation in order to obtain a major extended loop in each phase in turn, but the STL guide will only permit them to carry out a maximum of six break operations at the Test Duty 5 level. During development tests, it is good practice to arrange for clipped amplified voltage records across each phase to look for possible problems with the test circuit breaker contacts when experiencing peak fault currents.

13.3.4.3 Short time current test

A short time current test is carried out to confirm that the circuit breaker being tested is able to withstand safely its rated peak and rms short-circuit currents. The contact loading of the circuit breaker must be high enough to prevent burning or welding at the contact interface, and the thermal capacity of the current carrying components must be such that no problems will materialise on the short time current test. Once again, it is good practice to employ amplified arc voltage traces when conducting development tests, as these will show up potential problems as the test generator excitation is progressively increased. An oscillogram of a short time current test is shown in Figure 13.15.

The electromagnetic forces that are produced by Test Duty 5 and the first peak of the short time current test may cause damage to brittle insulation on conductors. It is therefore, good practice to thoroughly examine this insulation at stages during the development tests.

13.3.4.4 Other switching duties

Other switching duties may be proved by type tests if the test circuit breaker is to be assigned specific ratings. These include:

- small inductive current switching;
- single capacitor bank switching;
- back-to-back capacitor switching.

All of the above are discussed in Chapter 6.

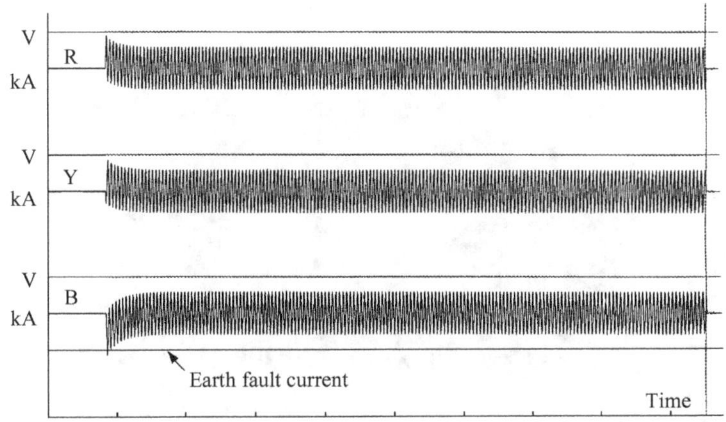

Figure 13.15 *Oscillogram of a short time current test (courtesy of the KEMA Laboratories, Netherlands)*

13.3.4.5 Internal fault tests

The internal fault tests are designed to prove that personnel will not be injured in the event of an arcing fault within switchgear. These tests are the most expensive that a manufacturer has to face because prototypes are typically five to ten times the cost of production units, and these tests effectively scrap the units being tested. At the outset of the tests, there is a degree of technical speculation and, under what is to some extent a trial-and-error process, several repeat tests may be necessary. This is because the effect of the rapid pressure rise within flat-sided cubicle constructions and arc burn-through time cannot be accurately predicted.

The Electricity Association standard EATS41-36 'Distribution switchgear for service up to 36 kV' specifies that internal fault tests are mandatory. The tests are also detailed in IEC60298, CENELEC prEN 50179, HN64-S40, S41 and S42 in France and DIN VDE 0670 in Germany.

EATS41-36 specifies that the magnitude of the current to be used for the tests is to be equal to the rated symmetrical breaking capacity of the switchgear being tested, and the duration of the current is specified as one second. In all other respects, the tests are to be in line with the requirements of IEC60298.

For these tests, fuse wire is used to initiate an internal fault separately within the busbar, circuit breaker and cable box spaces.

The positions where personnel are likely to be in the course of their normal duties are specified in the standards, and are simulated by cotton squares mounted on steel frames. These cotton squares, known as 'indicators', must not be ignited or pierced by ejected debris. Figure 13.16 shows a circuit breaker with a framework erected and fitted with the cotton indicators in their frames.

The tests can be quite dramatic to witness as considerable noise, smoke and flame is produced. Quite often the smoke will prevent observation of the test unit for a time. An impression of this test can be seen in Figure 13.17.

Figure 13.16 A circuit breaker with frames and indicators (courtesy of the KEMA Laboratories, Netherlands)

Figure 13.17 A circuit breaker on internal fault test (courtesy of the KEMA Laboratories, Netherlands)

The inclusion of 'blast-flaps' in the switchgear enclosure is the usual solution that is applied to dissipate the pressure build-up which will occur if an internal fault is to be allowed to burn for the full rated duration. The blast flap is usually a hinged cover which opens within about 10 ms of the start of the fault. The vent area provided by the blast-flap can be quite large and for smaller fault levels may represent the point in time of the generated peak pressure. However, at higher fault levels, the internal pressure may continue to increase after the blast-flap has opened, but at a lower rate. Development engineers have found that with some equipment, especially when tested for a fault duration of greater than 1 s, certain areas of the design will

Table 13.1 Test criteria for various levels of protection

Criteria no.	Assessment of test
1	Doors and covers do not open
2	Hazardous parts do not fly off
3	No burn-through of external surfaces
4	Vertical indicators do not ignite
5	Horizontal indicators do not ignite
6	All earthing connections still intact
Proposed 7	Adjacent compartments can be returned to service after cleaning

required a second metal skin to prevent arc burn-through, regardless of the current magnitude. Some designs even have a built-in feature that is intended to act as an arc root point in order to control the arc's position. The development tests also quite often show that some reinforcement of the construction is necessary, particularly in the area of doors, removable covers and chambers having large flat sides. Table 13.1 shows the test criteria for various levels of protection.

Internal fault test specifications fail to recognise that the protection provided is limited to blast and heat, which, while a big advance on no protection at all, does not protect the operators from other phenomena. For example, hearing damage may be caused by the associated very loud sound, and no protection for personnel is provided against exposure to arced SF_6 if this is vented within a substation.

Those who specify switchgear should note that the internal dimensions of substations will influence the internal fault test results. For example, the blast from faulted equipment could be reflected back, perhaps towards an operator, so the height of the substation ceiling is important. The specifier should study the type test report and ensure that the dimensions of the proposed location are equivalent to that used for the tests. Consideration should also be given to the substation materials to ensure that they can withstand hot gases and flames being issued from faulted equipment. Finally, there are ways of meeting the requirements for operator safety in the event of an internal fault by not letting it continue for the full rated duration. These are novel solutions which are not yet recognised by existing standards [19]. Some of these solutions are:

- Fitting a robust auxiliary switch to the blast flaps, so that the incoming feeder circuit breaker will be tripped in the event of the blast flap opening. Correct operation of this system can be verified by a simple test at any convenient time in the life of the switchgear.
- The inclusion of a light detector in the zone to be protected which will initiate a signal to a relay where it will stored, even if the light detector is destroyed by the fault. A second signal would be sent simultaneously from a current transformer in the feeder circuit breaker, and when both signals are present, the relay will initiate tripping of the incoming feeder circuit breaker.

- The third method, used in France, is similar to the previous one, but instead of using a light detector, it measures the pressure rise within a faulted compartment.
- A fourth method, used in Switzerland, is similar to the previous two, but uses sound instead of light or pressure to detect an internal fault. However, this method cannot identify an individual faulted chamber.
- Finally, there is another novel method that is worthy of note. This was introduced by Driescher in Germany and uses deflection of the walls of an SF_6 compartment to trigger the closure of a three-phase earth switch on the incoming side to effectively short out the fault. It is obviously important that any protection which uses a separate switch to short out a fault must be capable of having its mechanical operation checked during the life of the equipment.

13.3.5 Environmental tests

The environment can influence the performance of switchgear in different ways, and a series of tests must be carried out to ensure that the switchgear is immune to adverse environmental influences and is able to perform correctly in all specified conditions. The aspects of performance that have to be checked are:

(a) gas tightness;
(b) operating times and speeds;
(c) water ingress; and
(d) icing.

13.3.5.1 Gas tightness

Halogen gas leakage detectors are sensitive, calibrated measuring devices, and manufacturers use them to measure their success in achieving their target of zero gas leakage. In order to verify a virtually zero gas leakage rate, measurements must be made under high and low temperature conditions, as well as before, during and after mechanical operations tests. The reason for this is that gas sealing materials will expand and contract with changes in temperature at a faster rate than metals used in parts, such as shaft and covers. In addition, certain types of seals will lose their elasticity or even become brittle at low temperatures.

If there is going to be a gas leakage problem, it will occur in one of three conditions: the first condition is when the switchgear is being returned to normal ambient temperature after being held at its minimum rated temperature for sufficient time for all parts to reach that minimum temperature. Due to loss of elasticity, shaft and cover seals may lose their flexibility and not be able to maintain a sealing pressure on their associated metal parts when they start to warm up and expand. The second condition is at high temperatures, when the seals may have become semi-plastic and, therefore, lost their ability to remain effective.

Finally, dynamic seals may become ineffective due to wear or deformation caused by repeated mechanical operations.

For reference, various types of leak detection methods and the sensitivity of each, are given in a tabular format in IEC 60694 Annex E, which also gives an example of summation tests for interconnected compartments.

13.3.5.2 Operating times and speeds

To be certain of correct co-ordination with protection schemes and with other equipment on the distribution system, it is essential that the opening, closing and contact travel speeds are measured at both high and low temperature as well as periodically during the mechanical operations tests. These times and speeds need to be essentially the same at both the start and the end of the mechanical operations tests, and any variation at high or low temperature should be noted in the type test report.

13.3.5.3 Water ingress

Water entering switchgear will, almost certainly, have disastrous consequences. All outdoor switchgear is, therefore, required to be subjected to water ingress tests. These tests are detailed in Annex C of IEC 60694, which specifies water spray nozzle positions and rates of water flow.

The tests are deemed to be successful if:

(1) Apart from that on external insulation, no water is visible on the insulation of the main and auxiliary circuits.
(2) No water is visible on any internal electrical components or mechanism.
(3) No significant accumulation of water shall be retained by the structure, as it could cause corrosion.

Clearly, if water does enter and accumulate within the structure, the definition of the word 'significant' in clause 3 of Annex C would be of paramount importance. A prudent engineer would avoid any discussion by not allowing any water to enter a unit.

13.3.5.4 Icing tests

Tests need to be carried out to confirm that ice accretion in and around external drive shafts and handles will not prevent correct operation. This can be a particular danger with open type, pole-top switchgear.

The international standard IEC 60129 specifies how the ice should be formed and the tolerances on its thickness, which is measured on a test bar. Once the ice has been formed, the switchgear being tested has to be held at a temperature of −7°C for at least 4 h. This is to ensure that all parts, and the ice, have assumed a constant temperature.

Having achieved a constant temperature, the unit must operate correctly on the first attempt to move to its final closed or opened position. The switchgear should not have sustained any damage that could later interfere with its electrical or mechanical performance. Switchgear designers should prevent a possible operational problem by arranging for the first part of the operation to have an ice-breaking feature, such as an internally expanding scissor type lever mechanism with a high mechanical advantage.

13.3.6 Electromagnetic compatibility tests

The interaction between switchgear and the circuit in which it is installed can, during switching, give rise to voltage spikes and high-frequency voltage oscillations which

Figure 13.18 A subterranean EMC test facility

have the potential to interfere with other equipment. It is important that steps are taken to prevent or minimise this electromagnetic interference. Generally, there are three ways in which possible interference can take place. These are via conduction, radiation and induction, and there are remedies that eliminate or minimise all three types.

Switchgear that uses electromechanical relays will clearly not be interfered with by external sources, but some newer forms of electronic auxiliary switches and alternatives to conventional current and voltage transformers are susceptible and need protection. Within current conventional switchgear, electronic control and protection relays are now widely used, and these could suffer maloperation if the electromagnetic field levels in their vicinity exceed the levels of protection that they have built into their construction. Short-circuit testing of switchgear fitted with protection and control relays can prove the combination, or will very quickly show that there is an interference problem.

National and International standards at present do not call for switchgear to comply with the EU Directive on Electromagnetic Compatibility (EMC), as they are currently considered to be part of a complex whole. However, operation and control relays do have specified levels of compatibility and they must be separately proven. The situation with regards to standards needs to be carefully watched as changes may be introduced. Some customer standards specify levels of EMC, and most manufacturers will have carried out tests to determine the characteristics of their switchgear. Tests are usually conducted within a dedicated test house, and an interesting one is shown in Figure 13.18.

The test facility shown in Figure 13.18 is housed within a disused salt mine in Cheshire, which was found to be an ideal environment for the elimination of extraneous electromagnetic interference.

Chapter 14

Product conformity, quality control and service problem resolution

As it is neither practicable, nor economically feasible to type test every production circuit breaker or switch, instead, manufacturers are confined to carrying out inspections, measurements and certain limited tests, on a routine basis, to confirm that every production unit is identical to the unit that was type tested.

14.1 Serial numbers

Central to the process of the assurance of product conformity is the allocation of serial numbers. All separately tested and inspected components, such as insulators, current and voltage transformers and vacuum interrupters, should be allocated a unique serial number which should be indelibly marked on the component. This will allow data including the date of test, together with details of the tests and inspections carried out, and the names of the personnel involved to be recorded for examination at a later date if necessary.

The final assembly of the individual circuit breaker, or switch, should also be allocated a unique serial number. The in-house test and inspection records for the circuit breaker, or switch, should also list the serial numbers of all numbered sub-components. Should a problem arise, records such as these would be essential to allow the size of a problem and the location of affected units, to be established.

While the make-up of the actual serial number is not important, so long as it is unique, a connection with the contract number or year of manufacture would be useful. Manufacturers often use one or the other of these pieces of information in their serial numbers.

14.2 Routine test

Part of the process of confirming that production units will be identical in performance to the unit that was proven by type tests is done by implementing the requirements of ISO 9000 quality assurance on each production unit.

Certain routine tests are specified as mandatory in the relevant standards, such as IEC62271–100, and must, therefore, be carried out on each unit. In addition, product knowledge gained by the manufacturer in the course of developing their product may dictate that certain additional tests or inspections need to be carried to ensure that the unit will function correctly in service and be able to meet the declared ratings.

The mandatory routine requirements are:

(i) power frequency voltage withstand tests on the main circuit;
(ii) voltage withstand tests on the control and auxiliary circuits;
(iii) measurement of the resistance of the main circuit;
(iv) mechanical operating tests;
(v) design and visual checks.

These are examined each in turn in the following.

14.2.1 Power frequency voltage withstand tests on the main circuit

This test is identical in its procedure, value and duration, to that of the power frequency type test described in Chapter 13. For example, 12 kV switchgear should be tested phase-to-phase and each phase to earth at 28 kV for 1 min. It is usually advisable to isolate any connected voltage transformers and surge suppressors from the main circuits for these tests.

14.2.2 Voltage withstand tests on the control and auxiliary circuits

This is a 2 kV voltage test to earth on the secondary wiring. Certain types of electronic equipment may have to be disconnected for these tests.

14.2.3 Measurement of the resistance of the main circuit

The resistance of each phase of the main circuit should be measured, using equipment such as a Ductor. Each resistance measurement should be noted and compared with the type test values and the specified design acceptance criteria.

14.2.4 Mechanical operating tests

Mechanical operating tests are particularly important and searching in that operating times and speeds are noted and compared with the rated values. These tests and measurements not only verify correct functionality but also, by noting the differences in operating times at minimum, nominal and maximum operating voltages, confirm that the correct trip and close coils are fitted. They also confirm, by measuring the motor charging time, that the correct mechanism charging motor or solenoid is fitted.

A typical routine mechanical operations record is shown in Figure 14.1.

The record traces shown in Figure 14.1 are as follows.

(1) The vertical dotted lines are time markers, usually shown at 10 ms intervals, with time increasing from left to right.

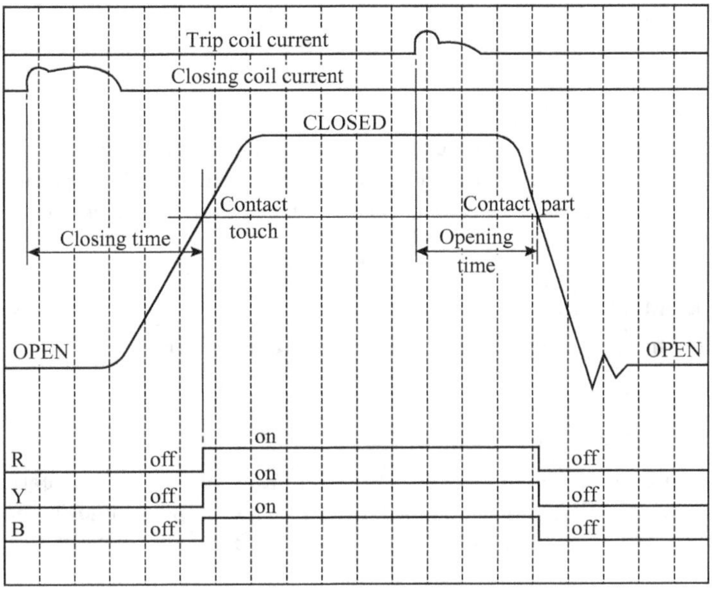

Figure 14.1 A typical mechanical operations record

(2) The top trace shows the current flowing through the trip coil.
(3) The next trace is similar, and shows the current flowing through the closing coil.
(4) The third trace from the top records the movement of the moving contact. This is usually obtained by attaching a linear transducer, or rotary potentiometer, to a mechanism component that moves with, and, therefore, mirrors the movement of the contacts.
(5) The bottom three traces measure the voltage across the main contacts and are usually obtained by applying a low voltage across the incoming and outgoing terminals of the switchgear. The purpose of these traces is to indicate when the contacts close and open.

The sequence of events shown in the record, starting at the left-hand edge begins with the main contacts in the OPEN position, confirmed by the bottom three traces. The second from the top trace then shows current starting to flow in the closing coil. After that, the moving contacts are shown to move to the CLOSED position; the bottom three traces indicating contact touch, which for finger type contacts, is before the main contacts reach the end of their travel.

The measurements made from the mechanical operations record are as follows.

14.2.4.1 The circuit breaker closing time

This is the time to contact touch from initiation of closing coil current. This will be measured with the trip coil voltage at its minimum, nominal and maximum rated value. It may be found that due to lack of contact simultaneity, there is a spread of closing times between the phases. In these circumstances, the spread in time must not be greater than that measured in the prototype during the short-circuit-type tests.

14.2.4.2 Contact closing speed

The contact closing speed is measured at the instant of contact touch.

When this is measured manually, it is normally taken as the slope of a straight line joining a point at about 5 per cent of the contact stroke before contact touch with the actual point of contact touch.

14.2.4.3 Contact stroke

The contact stroke is simply the linear distance between the OPEN and CLOSED positions.

14.2.4.4 Circuit breaker opening time

This is the time to the contacts parting from initiation of trip coil current. A spread in time between the first and last contacts to part is less important during an opening operation than when during a closing operation.

14.2.4.5 Contact opening speed

This can either be measured in a similar way to that used for measuring the contact closing speed, or measured from contact part to the point where the mechanism dashpot starts to slow down the contact speed. The method used should be the most appropriate for the type of interrupter used.

14.2.4.6 Contact travel

Full contact travel to the OPEN position should be confirmed by measurement. Contact travel records often show some apparent oscillation before the contacts come to rest in the OPEN position. Quite often this is not real and only reflects oscillations in the travel transducer drive linkage.

This concludes the review of the measurements usually made following mechanical operation tests.

14.3 Automatic routine test facilities

Some manufacturers have designed and installed specialised test facilities which will automatically carry out all of the routine tests described in Section 14.2.4. An example of such a test facility is shown in Figure 14.2.

The test facility shown in Figure 14.2 was designed to carry out routine tests on the loose circuit breaker portions of two types of withdrawable switchgear. The left-hand side was designed to accept a horizontally isolated circuit breaker and the right-hand side was designed to accept a vertically isolated circuit breaker. Circuit breakers were labelled with a bar code which was automatically read by the test facility when a circuit breaker was engaged. The test facility used a look-up database

Figure 14.2 An automatic routine test facility (courtesy of ALSTOM T&D Ltd)

which provided all of the information necessary for an automatic test procedure to be carried out under computer control.

This information included the circuit breaker voltage, current and short-circuit rating, the secondary voltage rating and whether or not the operating mechanism was fitted with a motor charging facility.

The tests, once started, would be automatic and unmanned. Should any measured value be outside the acceptability limits of the design, the test would automatically stop and the computer would print out a defect report. If all of the tests were successful, the computer would print out a routine test certificate, detailing all of the readings. It would also automatically send the serial number of the circuit breaker to the manufacturing facility, for information, telling them that the routine tests had been successfully completed. Finally, it would send all of the results to quality control so that they could look for any significant trend in the results.

The automatic routine test facility takes all human judgement out of the tests, making the tests repeatable regardless of any changes in the personnel involved. However, while extremely useful, these types of test facilities are complex and expensive to produce. An idea of the complexity is given by a view of the rear of the facility shown in Figure 14.3.

14.4 Design and visual checks

This is the final specified mandatory routine check. The inspector carrying out this check should be provided with a list of points to check. Some companies provide a list with photographs and drawings to help identify the features to be checked.

Figure 14.3 Rear view of the automatic test facility shown in Figure 14.2 (courtesy of ALSTOM T&D Ltd)

The inspector should sign the inspection sheet against each item checked. Some switchgear companies in Japan and the UK provide each person in an assembly team with a unique coloured felt tip pen. As each fastening is torqued to the specified value, a line is drawn across the edge of the fastening and its related component. This not only tells the inspector that the fastening has been correctly tightened, but also who carried out the work.

The inspection sheet should include a fill-in box to allow the inspector to record the serial number of the equipment and that of each separately numbered sub-component. This form should then be kept in the contract file for future reference. In addition, these serial numbers should be recorded on a database for quick tracabilty in the event of a later problem materialising.

14.5 Quality control

Much of what has been written in this chapter so far can be said to be part of the quality control procedure necessary to produce a trouble-free product. However, to be effective, quality control personnel should carry out quality audits to ensure that the system is being properly implemented. It may be that certain quality checks can be seen to be unnecessary and can be discontinued, whereas others may need to be expanded. Also, any trends coming out of the routine test results may indicate that additional tests or checks should be introduced.

The quality standard ISO 9000 requires that design review meetings be held, attended by all relevant departments, to review service experience.

14.6 Design review meetings

For design review meetings to be effective, manufacturers should ensure that a company culture exists that encourages the reporting and review of all problems, either internal or onsite, howsoever trivial. The procedure should be that whoever has a problem reported to them is responsible for completing a complaint form and forwarding this to the quality assurance department. Each form should then be given a unique number and copied to all relevant department heads. These forms are used as the basis for the design review meetings at which each problem is discussed and corrective actions agreed to prevent a re-occurrence of the problem.

14.7 Service problem resolution

It is difficult to visualise a problem which is not the result of someone not doing their job properly. The problem may lie with the design, in terms of the suitability of components, with the manufacturer in terms of component quality, or with the user who has not installed the equipment correctly nor understood the limitations of the switchgear that is installed. The problem could be due to a higher than anticipated fault level at the location, or due to non-rated switching being carried out, for example, back-to-back capacitor switching. However, experience has shown that most problems lie with the manufacturer.

Problems with switchgear in service are expensive for both the manufacturer and the user. The manufacturer, who is anxious to protect his reputation, may incur costs to manufacture components and carry out corrective action on site, which could be anywhere in the world. A factor of ten is said to be involved in the expenditure incurred at each stage of a quality problem. If the expenditure costs are X for a problem found at the component manufacture stage, then it will be $10X$ if the components complete the assembly stage and $100X$ if the assemblies are delivered to customer and installed on site.

The user's expenditure will depend upon the nature of the problem. It may involve having to institute switching restrictions and to do this he may need to divert labour from other tasks and possibly have to pay for out of hours working. He may also face loss of production or flexibility in terms of the system network. It is, therefore, essential that the problem is minimised and eliminated as soon as possible.

14.8 Minimising the problem

It is often possible to limit the size of a service problem when the nature and cause of the problem is understood, especially when the problem involves a component having a serial number. The stages to be followed in an investigation are shown in Figure 14.4.

Take the hypothetical example of a mechanically stressed moulded insulator failing after only 50 operations in service. It is no good the manufacturer saying that type

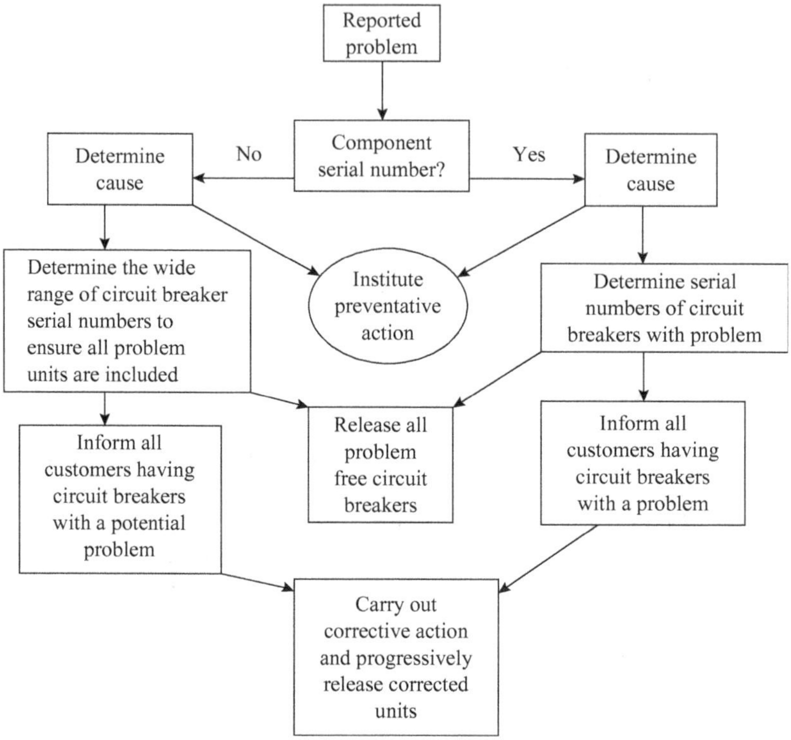

Figure 14.4 Stages in a service problem investigation

tests of 10,000 operating cycles were carried out on the prototype. Either the forces acting on the insulator were, for some reason, extremely high or the insulator was, for a reason yet to be determined, very weak.

The investigation would start by establishing the serial numbers of the moulded insulator which failed. This would, in turn, show who actually manufactured the mouldings and when. A subsequent investigation with the moulding manufacturer may establish from, say, a glass transition temperature test that the mouldings were not properly cured in the mould or post-moulding fixture. Possibly a heater in the moulding tool could have failed and a check on this would confirm this to be the case. All moulded insulators produced using that particular mould would, therefore, be suspect. The manufacturer's database would give the serial numbers of the circuit breakers fitted with the suspect moulding together with the names of the affected customers. Four things would then happen:

- All customers having suspect circuit breakers would be informed.
- All circuit breakers not involved would be cleared for normal operation.
- Corrective action would start.
- Preventative action would be instituted.

This example is, of course, hypothetical, and only used to demonstrate how a potentially large problem can be contained and resolved. In practice, a routine glass transition temperature test is usually carried out each week to identify any potential mould-curing problem before mouldings are sent to the assembly area of the equipment manufacturer.

Problem limitation becomes more difficult if parts not having serial numbers fail due to a latent defect. The investigators then have to look for another way of reducing the numbers involved. For example, a spring fails mechanically and the cause of the spring failure is established as a process failure at the sub-contractor's works. The spring manufacturer involved may be identified by the finishing colour applied to the springs if this colour is unique to one manufacturer. For example, if the failed spring was gold in colour, switchgear having springs in silver or black could be examined and released from any restriction provided the switchgear manufacturer issued clear instructions to customers on what to look for.

However, the situation becomes more difficult if a number of sub-contractors provide identical indistinguishable springs. This problem is compounded if all of the springs were held in storage bins and not used on a first-in, first-out basis. The records of goods inwards stores would give the supply dates of each manufacturer's springs, and in-house records would show the date when the springs were last out of stock in the assembly area. This may be the only cut-off date available to the manufacturer and so the list of suspect circuit breakers involved would be larger as it would include many circuit breakers that were, in fact, trouble free.

Chapter 15
Cost of ownership

The cost of switchgear ownership can be very much greater than simply the initial capital cost of purchasing the switchgear. Some accountants are only interested in the expenditure within a current financial year. However, it could be argued that this first cost will only be a small percentage of the total cost of ownership over the lifetime of the switchgear.

The true cost of ownership should include the cost of the substation, as different types of switchgear will occupy different volumes of substation space and, therefore, influence the overall substation cost. In addition, costs will be incurred for the erection, installation and commissioning of the switchgear. Once the switchgear is installed, other costs will be incurred, such as maintenance, labour, materials and outage time. Finally, disposal at the end of useful life must be considered. For example, SF_6 switchgear will incur disposal costs, as only specialist companies have the facilities necessary to safely dispose of the equipment without risk to personnel and without contaminating the atmosphere. Conversely, disposing of oil and vacuum switchgear at the end of their life will generally be self-financing or even yield a small profit. All of these costs combine to give the true lifetime cost of ownership.

There are many ways of arriving at the true cost of ownership, but the following is one that is easy to comprehend and allows the costs to be expressed in today's value of currency.

Given that:

The substation land and building costs total $= S$
The initial cost of the switchgear $= A$
The life of the switchgear (years) $= L$
The maintenance interval (years) $= M$
The cost of maintenance, including the cost of spares and transport to and from site $= C$
The number of maintenances sessions over the life of the switchgear $= n$
The bank interest rate (%) $= I$
End of life disposal cost $= D$

Then, over the lifetime of the switchgear:

The basic cost of the substation and the purchase and installation of the switchgear will be $= (S + A) + [I \times L(S + A)]$
$= (S + A)(1 + I \times L)$

The cost of the first maintenance $= C + (L - M)C \times I$
$= C[1 + I(L - M)]$

Similarly, the cost of the second maintenance $= C[1 + I(L - 2M)]$
and so on, until $L - (n \times M) = 0$.

The full expression for the cost of ownership will be:

$$\text{Total cost} = (S+A)(1+I \times L)+C[1+I(L-M)]+C[1+I(L-2M)] \text{ etc.}+D$$

Example 15.1

Assume that the cost of the purchase of a substation building, the internal SF_6 switchgear together with its installation and commissioning costs is 100,000 units of currency.

The switchgear has a maintenance interval of 5 years and each maintenance costs 2000 units of currency. Given that the life of the installation is 40 years, the bank interest rate is 4 per cent and the end-of-life disposal is 10 per cent of the original installation cost, what will be the total cost of ownership?

Purchase, installation and commissioning costs $= (S + A) \times (1 + I \times L)$
$= 100{,}000[1 + (40 \times 4/100)]$
$= 260{,}000$.

Cost of the first maintenance $= C[1 + I(L - M)]$
$= 2000[1 + 4/100(40 - 5)]$
$= 4800$

Cost of the second maintenance $= C[1 + I(L - 2M)]$
$= 2000[1 + 4/100(40 - 10)]$
$= 4400$

And similarly
Cost of the third maintenance = 4000
Cost of the fourth maintenance = 3600
Cost of the fifth maintenance = 3200
Cost of the sixth maintenance = 2800
Cost of the seventh and final maintenance = 2400
Therefore the total cost of maintenance = 22,800
The cost of end-of-life disposal $= 10/100 \times 260{,}000$
$= 26{,}000$
Therefore, the total cost of ownership $= 260{,}000 + 22{,}800 + 26{,}000$
$= 308{,}800$.

Table 15.1 Comparative total cost of ownership

	SF_6 switchgear		Vacuum switchgear	
Substation and switchgear costs	100,000	100,000	100,000	100,000
Life of the switchgear	40	40	40	40
Maintenance interval (years)	5	5	10	10
Maintenance cost per visit	2000	2000	1000	1000
Bank interest rate	4	6	4	6
End-of-life disposal cost	26,000	26,000	0	0
Total cost of ownership	308,800	396,800	265,000	346,000

Having a model of cost build-up, it is possible to see the influence of the bank interest rate, the maintenance interval, the maintenance cost together with the disposal cost on the total cost of ownership.

If we now re-calculate the figures for the previous example, only changing the bank interest rate to 6 per cent, we can see the influence of interest the rate on the total cost of ownership.

Finally, we can rework the figures for interest rates of 4 per cent and then 6 per cent, assuming that the maintenance interval was increased to 10 years and maintenance cost reduced to 1000 units of currency, with no end-of-life disposal costs. This reduction in maintenance interval, together with the elimination of the end-of-life disposal costs illustrates the difference in total cost of ownership if vacuum switchgear was used in place of SF_6 switchgear. The initial purchase and commissioning costs have been maintained at the same level so that a comparison of total costs will give an idea of the relative costs of the different options. The results of these calculations are shown in Table 15.1.

The calculations show that reduced maintenance costs and increased maintenance intervals, together with the avoidance of end-of-life disposal costs, can significantly reduce the cost of ownership. In the example shown, this would mean that a premium of 40,000 units of currency could be paid for the vacuum switchgear option while still achieving an overall reduction in the total cost of ownership.

Chapter 16

The future

Forecasting the future is not without its hazards. History is littered with the failed prophecies of those who dared to forecast what was going to happen. The cause of this is that something completely unexpected crops up that turns the forecast on its head. So, being mindful of the hazards, these predictions are based upon the best information to-hand. Steering the ship by looking at its wake is far from ideal but does give an idea of where things will probably go.

Change is brought about by drivers which are both technical and economic. The main drivers for change are shown in Figure 16.1.

Let us now look at the way in which each of these drivers brings about change.

16.1 Technology

Discreet enquiries have been made within the personal network of universities and industry. These suggest that while the influence of new materials and shapes is being

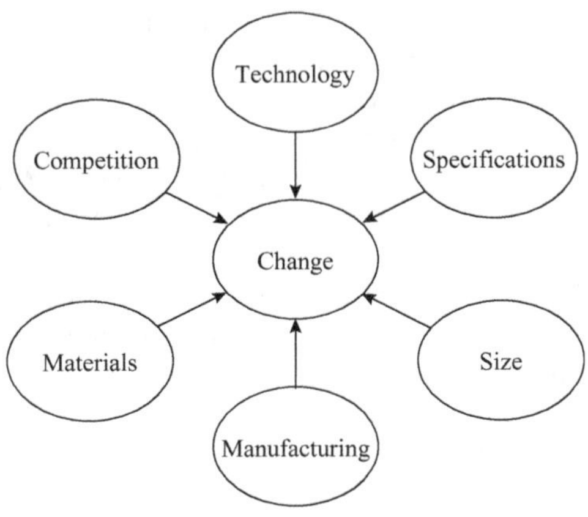

Figure 16.1 The drivers of change

widely explored in an attempt to reduce costs, as shown in Chapter 9 (Figure 9.5), there are no new technologies on the horizon that would compete with vacuum and SF_6.

There was some hope that liquid sodium devices would be a contender for a place in distribution switchgear, but it was only able to find an application within special low-voltage fuses.

As a significant proportion of problems with distribution switchgear involve parts that move, solid state switching held out the prospect of a brand new circuit breaker having no moving parts and no contact erosion. However, the ratings available of back-to-back solid state devices such as Triacs were such that a large number of devices had to be used connected in series and parallel in order to reach usable distribution voltages and current ratings. Voltage sharing of series connected devices needed shorting capacitors and the heat produced by the forward voltage drop meant that oil cooling with circulation pumps would be necessary. A costing exercise was carried out and this showed that a solid state circuit breaker would cost about ten times the price of a conventional distribution circuit breaker having the lowest usable rating.

If something like fuel cells reached the stage in development where sufficient power for a household could be contained in a box the size of a suitcase. Further, if we suppose that this box only needed to be replaced once a year, then there would be no need for power stations, overhead lines, buried cables, transformers or primary and secondary switchgear. One cannot imagine a bigger change than that, however, if this box were to explode everyone would want to be at least 100 miles away!

16.2 Specifications

Established National and International Specifications, like those listed in Chapter 18, have maintenance teams assigned to them, whose function is to regularly review their contents to ensure their continuous relevance in the light of experience. From time to time, changes are introduced to protect against possible weaknesses and to recognise changes in technology. These maintenance teams are drawn from experts within the area of the users, manufacturers and Health and Safety personnel. Many of the changes are relatively minor, but occasionally a major change is introduced. All manufacturers are informed of pending changes by their trade associations, indeed, many of the experts from the manufacturing side would have been nominated by their trade association to represent other members. The users will be similarly informed.

Examples from the past of major changes include the non-acceptance of direct manual operation of circuit breakers, and the introduction of a fault make operation before the break operation at test duty levels 1–4. Both of these changes were introduced as a result of catastrophic failures in service.

Revisions to standards can be found in the year book published by the authority issuing the standards so that manufacturers and users are made fully aware of all changes.

16.3 Competition

Competition is one of the great drivers of change and without competition very little would change. The story of the world beating a path to the door of the man who invented a better mousetrap is very true. Smaller, lighter, with higher ratings and functionality will always spur the competition to develop something even better. Figures 9.1, 9.2 and 9.5 illustrate this perfectly.

There is another aspect of background change brought about by competition within the electricity generation market and this concerns one of the effects of embedded generation. Whenever new embedded generation is introduced, it causes the short-circuit fault level in the local area to be increased. This often means that the increase in fault level puts it beyond the ratings of existing installed switchgear. The switchgear manufacturer can sometimes upgrade installed switchgear, in terms of short-circuit and normal current rating by changing elements within its construction to a proven higher rated variant. This may mean changing vacuum interrupters and mechanism springs. They are sometimes able to do this because the type of switchgear being considered may have already been proven for enhanced ratings in order to meet applications within the industrial sector rather than the distribution markets.

If, however, no enhanced rating is available within the considered switchgear range, then alternative types may well have to be sourced. The problems that will then have to be addressed include the design of the substation to accommodate the physical size of the replacement switchgear. Remembering that changes in substation dimensions from those dimensions used for internal arc fault tests could invalidate the test results.

16.4 Materials

Discoveries and developments in manufacturing, which offered benefits in size and cost, will bring about change. Within the sphere of distribution switchgear, insulation has seen very big changes. These are listed in chronological order in Chapter 7, starting with porcelain, slate and rubber, and progressing to the epoxy resin, glass filled epoxy and polyurethane resin found in modern distribution switchgear.

16.5 Manufacturing

Costs are taken out of products by analysing the time and labour content. New manufacturing processes can often lead to a saving in both. A good example of how this driver brought about change can be found in the evolution of pole mounted autoreclosers.

The original oil-filled units, which interrupted current using arc control pots, as shown in Chapter 2 (Figure 2.3), used the switch oil within its sequence control assembly. The reset dashpot of this assembly had to take between 60 and 90 s to reset, and only manual polishing of the dashpot bore could yield the required accuracy. This made it very expensive to adjust and test and meant that the routine adjustment and test of one unit could often take up to one man-week to complete. Another feature of the oil autorecloser was that it had to be brought back to a central workshop for overhaul and maintenance. The change from oil to vacuum for current interruption and the change from hydraulic sequence control to electronic control meant that the routine test in an automatic test facility could be used, bringing the time down to less than 30 min per unit, and the autorecloser would be maintenance free.

16.6 Size

The cost of land for substations can be very large, even when subterranean substations are used. This is particularly true for distribution switchgear located within large cities. Small overall dimensions therefore can offer the manufacturer a significant edge over the competition. An example of just how compact modern switchgear can be is shown in Chapter 10 (Figure 10.16(b)).

16.7 Manufacturing base

So far we have discussed the main drivers for change, but there is another growing influence that may have an impact on the design of distribution switchgear. This is the reduction in the total of individual manufacturers that has taken place over the last 20 years. This reduction in numbers has come about by take-overs and mergers. Ultimately, there may only be about six major world-class manufacturers. Through time, within these larger manufacturers, centralisation of R&D can be expected. This is a two-edged sword in that, on the one hand, these R&D departments can expect to be larger, employing top class engineers, however, a reduction in variety can also be expected in order to maximise the efficiency of the manufacturing units. What we may be left with is a high-quality product with the flexibility to meet all requirements on the basis of 'one size fits all'. In other words, there is a danger that the reduction in competition numbers may reduce the ingenuity in individual designs that have been seen in the past.

16.8 The shape of things to come

Having looked at the drivers for change, and at some of the changes that have taken place, we can now speculate on what is possibly going to be produced in the future. It is understood that this exercise can be likened to steering the ship by looking at its wake, but at the risk of running onto a reef, it should be possible to get an idea of what will be revealed in the coming years. The following are nine suggestions:

- The only technology to be used for fault interruption will be vacuum and SF_6 interrupters and for secondary switchgear, high-voltage fuse-links. A reduction in size for each will be realised.

- It is likely that on environmental and end-of-life cost grounds, SF_6 gas will take a smaller share of the market.
- When sufficient service experience has built up, there may be a change from SF_6 gas to solid insulation.
- There will be a continued growth in the use of magnetic actuators in place of conventional spring operating mechanisms.
- In the future, most units will be remotely controlled, to reduce operating man-hour costs and outage time.
- Remote monitoring of circuit conditions such as voltage current and power will become the norm.
- Because of the advantages that this gives to the designer and manufacturer, the use of alternatives to conventional current and voltage transformers will grow.
- The use of microprocessors in both protection and control has resulted in some blurring of the distinction between these functions, and it is expected that they will be unified in the near future.
- Finally, knowledge-based centralised protection linked to units in the field by secure communications is likely to take place. This will give the protection a global view of networks and allow automatic circuit reconfiguration in the event of a system fault.

Chapter 17
Further reading

The following lists sources of information which will assist the reader to search and explore aspects of switchgear in greater depth.

17.1 Books

'The switchgear handbook, vol. 1 – apparatus' (Sir Isaac Pitman & Sons Ltd., London, 1953)

VLADISLAV, Z.: 'High voltage circuit breakers' (Constable & Co. Ltd., Prague, 1957)

LYTHALL, R. L.: 'The J & P switchgear book' (Peter Peregrinus Ltd, London, 1982)

FLURSCHEIM, C. J.: 'Power circuit breaker theory and design' (Peter Peregrinus Ltd., London, 1982 revised edition)

BLOWER, R. W.: 'Distribution switchgear' (Collins, London, 1986)

RYAN, H. M., and JONES, G. R.: 'SF_6 switchgear', IEE Power series book, vol. 10, (Peter Peregrinus Ltd, London on behalf of the IEE, 1989)

GREENWOOD, A.: 'Electrical transients in power systems' (John Willey and Sons, Inc., Chichester, 1991, 2nd edn.)

GREENWOOD, A.: 'Vacuum switchgear', IEE Power series book, vol. 18, (IEE, London, 1994)

WRIGHT, A., and NEWBERY, P. G.: 'Electric fuses', IEE Power series book, vol. 2, (IEE, London, 1994).

17.2 Papers and published articles

SLEPIAN, J.: 'Extinction of an a.c. arc', *Transactions of the American Institute of Electrical Engineers*, 1928, **47**

CASSIE, A. M.: 'A new theory of arc interruption and circuit severity'. CIGRE, France, 1939

BALTENSPERGER, P.: 'Overvoltages due to the interruption of small inductive currents'. CIGRE, France, 1950

BLOWER, R. W.: 'Factors influencing the interruption in electric power supply systems'. Ferguson Palin Ltd. publication 1045/290, 1954

REECE, M. P.: 'The vacuum switch – parts 1 and 2', *Proc. IEE*, 1963, **110**, pp. 793–811

CHRISP, G. F.: 'Electromagnetic tripping devices'. UK Patent 1,236,916, 1964

MITCHELL, G. R.: 'High current vacuum arcs', *Proc. IEE*, 1970, **117**, pp. 2327–2332

GREENWOOD, A. N., KURTZ, D. R., and SOFIANEK, J. C.: 'A guide to the application of vacuum circuit breakers', *IEEE Trans.*, 1971, **PAS-90**, pp. 1589–1597

KIMBLIN, C. W., *et al.*: 'Interruption ability of vacuum interrupters subjected to an axial magnetic field', *Proc. IEE*, 1972, **119**(12), pp. 1754–1758

MORIMIYA, O., *et al.*: 'High current vacuum arcs stabilised by axial magnetic fields', *Trans. IEEE*, 1973, **PAS-92**, pp. 1723–1732

BLOWER, R. W., CORNICK, K. J., and REECE, M. P.: 'The use of vacuum switchgear for the control of motors and transformers in industrial systems'. IEE international conference of *Developments in distribution switchgear*, 1978

STEWART, J. S.: 'SF_6 circuit breaker design and performance', *IEE Electronics & Power Journal*, **February** 1979, pp. 121–126

STEWART, J. S.: 'An autorecloser with microprocessor control for overhead line distribution', *IEE Electronics & Power Journal*, June 1984, pp. 469–472

PARRY, J.: 'Trends in modern switchgear design'. IEE Symposium, 1984

PARRY, J.: 'Development of SF_6 switchgear incorporating rotating arc circuit breakers', *IEE Electronics & Power Journal*, August 1986

FALKINGHAM, L. T.: 'Recent advances in vacuum interrupter design', *GEC Review*, 1986, **2**(3)

CORNICK, K. J.: 'Current chopping performance of distribution circuit breakers'. IEEE winter meeting, London, 1986

STEWART, J. S.: 'Further developments in distribution switchgear for overhead networks'. IEE colloquium on *Improving supply security on 11 kV overhead networks*, London, digest 1988/134, December 1988

STEWART, J. S.: 'Oil-less switchgear'. IEE Hong Kong centre lecture, March 1990

STEWART, J. S.: 'Meeting the objectives for operator safety'. IEE colloquium on *Risk reduction: internal faults in T&D switchgear*, Nottingham, April 1997

McKEAN, B., and REUBER, C.: 'Magnets & vacuum – the perfect match'. IEE 5th international conference on *'Trends in distribution switchgear'*, conference publication 459, 1998

BONJEAN, M., *et al.*: 'An asymmetrical magnetic actuator for MV circuit breakers'. CIRED, 1999

STEWART, J. S.: 'Magnetic actuators applied to primary and secondary switchgear'. CIRED, 1999

STEWART, J. S.: 'Primary switchgear', *IEE Power Engineering Journal*, 2000, **14**(6), pp. 264–269

LEEWERKE, R. P., *et al.*: 'Developments in ring main unit design for improved MV network performance', *IEE Power Engineering Journal*, 2000, **14**(6), pp. 270–277

Chapter 18

National, International and customer Specifications

As specifications are under continuous review, everyone associated with the design, manufacture, application and operation of switchgear should ensure that they consult with the latest edition of a relevant specification. They can verify this by checking the yearbook of the standards organisation. If they do not do this, then there is a danger that they will not comply with tenders or standard operation instructions and may put themselves into an illegal situation. The following standards are listed for guidance only and the yearbook should be consulted to ensure that the copy being used is up to date.

IEC 73	Basic and safety principles for man–machine interface, marking and identification. Coding principles for indication devices and actuators (6th edition 2002)
IEC 417	Graphical symbols for use on equipment (superseded by IEC60417 parts 1 and 2, superseded date 01/09/98)
IEC60417 part 1	Graphical symbols for use on equipment – part 1: overview and application (3rd edition 2002)
IEC60417 part 2	Graphical symbols for use on equipment – part 2: symbol originals (1st edition 1998, Amendment 2-2002)
IEC 815	Guide for the selection of insulators in respect of polluted conditions (1st edition 1986)
IEC 870	Tele-control equipment and systems
IEC 1330	High-voltage/low-voltage prefabricated substations (1st edition 1995)
IEC 60044 PTI	Instrument transformers – Current transformers (1st edition 1996, amendment 1-2000)
IEC 60056	High-voltage alternating current circuit-breakers (superseded by IEC62271-100, superseded date 01/05/01)

IEC 62271-100	High-voltage switchgear and controlgear – part 100: high-voltage alternating current circuit breakers (1st edition 2001, amendment 1-2002)
IEC 60129	Alternating current disconnecters and earthing switches (superseded by IEC62271-102 on 19/12/01)
IEC 62271-102	High-voltage switchgear and controlgear – part 102: high-voltage alternating current disconnectors and earthing switches (1st edition 2001, corrigendum 1-2002)
IEC 60186	Voltage transformers (2nd edition 1987, amendment 2-1995)
IEC 60265-1	High-voltage switches – switches for rated voltages above 1 kV and less than 52 kV (3rd edition 1998, corrigendum 1-2000)
IEC 60298	A.c. metal-enclosed switchgear and controlgear for rated voltages above 1 kV and up to and including 52 kV (3rd edition 1990, corrigendum 2-1998)
IEC 60420	High-voltage alternating current switch-fuse combinations (superseded by IEC62271-105 on 22/08/02)
IEC62271-105	High-voltage switchgear and controlgear – part 105: alternating current switch-fuse combination (1st edition 2002)
IEC 60480	Guide to checking of SF_6 taken from electrical equipment (1st edition 1974)
IEC 60529	Degrees of protection provided by enclosures (IP code) (consolidated edition 2001)
IEC 60617	Graphical symbols for diagrams
IEC 60694	Common specifications for high-voltage switchgear and controlgear standards (consolidated edition 2002)
IEC 61243	Live working-voltage detectors
IEC 61634	High-voltage switchgear and controlgear – use and handling of SF_6 in high-voltage switchgear and controlgear (1st edition 1995)
IEC 61958	High-voltage prefabricated switchgear and controlgear assemblies – voltage presence indicating systems (1st edition 2000)
IEV 441	International electrotechnical vocabulary – chapter 441: switchgear, controlgear and fuses (1990 version)
ISO 9000	Quality management systems – fundamental and vocabulary (2nd edition 2000)
BS EN 60168	Tests on indoor and outdoor post-insulators of ceramic material or glass for systems with nominal voltages greater than 1 kV (1995 version, amendment 13134-2001)
BS 148	Specification for unused and reclaimed mineral insulating oil for transformers and switchgear (version 1998)

BS 3SIC	Specification for colours for identification, coding and special purposes.
BS 2045	Preferred numbers (Associated Standards ISO 3, ISO 17 and ISO 497) (version 1965)
BS 2562	Specification for cable boxes for transformers and reactors (version 1979)
BS 2692	Fuses for voltages exceeding 1000 V a.c. (Associated Standard IEC 60282)
BS2874	Specification for copper and copper alloy rods and sections (other than forging stock)
BS 4608	Specification for copper for electrical purposes (rolled strip, sheet and foil) (superseded by BS EN 13599 on 19/09/2002)
BS EN 13599	Copper and copper alloys – copper plate, sheet and strip for electrical purposes (version 2002)
BS 5207	Specification for sulphur hexafluoride for electrical equipment (Associated Standard IEC 376) (version 1975)
BS 5559	Specification for identification of apparatus terminals and general rules for a uniform system of terminal marking, using an alphanumeric notion (Associated Standards EN 60445, IEC 445) (superseded by BS EN 60445 on 15/07/2000)
BS EN 60445	Basic and safety principles for man–machine interface, marking and identification of equipment terminals and terminations of certain designated conductors including general rules for an alphanumeric system (version 2000)
BS 5775	Specification for quantities, units and symbols (Associated Standard ISO 31)
BS 6121-1	Mechanical cable glands – specification for metallic glands (version 1989, partially superseded by BS EN 50262)
BS EN 50262	Metric cable glands for electrical installations (version 1999, amendment 13524-2002)
BS 6423	Code of practice for maintenance of electrical switchgear and controlgear for voltages up to and including 650 V (version 1983, amendment 6812-1992)
BS 6553	Guide for selection of fuse-links of high-voltage fuses for transformer circuit applications (Associated Standard IEC 60787) (version 1984, amendment 6685-1991)
BS 6626	Code of practice for maintenance of electrical switchgear and controlgear for voltages above 650 V and up to and including 36 kV (version 1985, amendment 6813-1991)

BS 7198	Hydraulic fluid power quick-action couplings (Associated standard ISO 7241)
BS 7354	Codes of practice for design of high-voltage open-terminal substations (version 1990, amendment 7160-1992)
BS 7735	Guide to loading guide of oil-immersed power transformers (version 1994)
ANSI/IEEE C37.60	American National Standard requirements for overhead, pad-mounted, dry-vault and submersible automatic circuit autoreclosers and fault interrupters for a.c. systems, version 81 (revised 1992)
ANSI/IEEE C37.63	American National Standard requirements for overhead, pad-mounted, dry-vault and submersible automatic line sectionalisers for a.c. systems (version 1997)
EATS 12-8	The application of fuse-links to 11 kV/415 V and 6.6 kV/415 V underground distribution networks (issue 2-1986)
EATS 12-11	Indoor and outdoor cable boxes for switchgear (for service at nominal system voltages of 6.6, 11, and 33 kV) (issue 2, amendment 1)
EATS 35-1	Distribution transformers (from 16 kV A to 1000 kV A) (issue 4, amendment 1-1993)
EATS 35-15	Protection and measurement transformers for high-voltage distribution systems up to 36 kV
EATS 41-16	Apparatus terminations, conductor sizes and associated fittings (copper used in outdoor and indoor substations with outdoor equipment) (issue 2 – 1981)
EATS 41-18	Partial discharge testing of bushings, capacitors, instrument transformers and switchgear of rated voltage 7.2 kV–420 kV inclusive (issue 2 – 1995)
EATS 41-36	Distribution switchgear for service up to 36 kV (cable and overhead conductor connected) (issue 1 – 2000)
EATS 43-92	Conductor fittings for overhead lines (issue 2, amendment 1 – 1993)
EATS 43-95	Steelwork for overhead lines (issue 5, amendment 1 – 1993)
EATS 48-2	Fault passage indicators for 6.6 kV and 11 kV underground and overhead distribution systems (issue 1, amendment 1 – 1993)
EATS 50-18	Design and application of ancillary electrical equipment (issue 2)
EATS 98-1	Surface preparation and coating systems for new plant and equipment (issue 2, amendment 1 – 1997)

BEBS S 12	Specification for standard numbering for small wiring for switchgear and transformers together with their associated relay and control panels (amendment 1)
Engineering Recommendation G 18	Operation of air break isolating switches
Engineering Recommendation S 15	Standard schematic diagrams

The Construction Design and Management Regulations 1994
The Electricity at Work Regulations 1989

References

1 CASSIE, A. M.: 'A new series of rupture and circuit severity'. CIGRE 1939. Paper 102

2 SLEPIAN, J.: 'Extinction of an AC arc', *Transactions of the American IEE*, 1928, **47**, p. 1398

3 TRENCHAM, H.: 'Circuit breaking' (Butterworth Scientific Publications, 1953)

4 STEWART, J. S.: 'Oil-less switchgear'. IEE Hong Kong centre lecture, March 1990

5 GREENWOOD, A.: 'Vacuum switchgear'. IEE Power series book (1994)

6 STEWART, J. S.: 'SF_6 circuit breaker design and performance', *IEE Electronics & Power Journal*, February 1979

7 PARRY, J.: IEE symposium on *Trends in modern switchgear design*, 1984, pp. 2.1–2.6

8 Kali und Steinsalz, **3** (10), 1963

9 ETZ Supplement 3, 1966

10 SIMMS, J. R.: 'Operating mechanisms'. EA Technology Switchgear Technology training course notes

11 CHRISP, G. F.: 'Electromagnetic tripping devices' UK Patent 1,236,916

12 BONJEAN, M., *et al.*: 'An asymmetrical magnetic actuator for MV circuit breakers'. CIRED, 1999

13 STEWART, J. S.: 'Magnetic actuators applied to primary and secondary switchgear'. CIRED, 1999

14 JOHNSON, K. J. and DILKES, G.: 'An energy saving arrangement for magnetic actuators'. UK and Foreign patent applied for

15 LEEUWERKE, R. P., *et al.*: 'Developments in ring main unit design for improved MV network performance', *IEE Power Engineering Journal*, 2000, **14**(6)

16 LOMA, K.: 'Reduction risk: Criteria for choosing an appropriate design approach'. IEE colloquium on *Risk reduction*, 14 April 1997

17 STEWART, J. S.: 'An autorecloser with microprocessor control for overhead line distribution', *IEE Electronics & Power Journal*, **June** 1984

18 STEWART, J. S.: 'Primary switchgear', *Power Engineering Journal*, 2000, **14**(8)

19 STEWART, J. S.: 'Meeting the objectives for operator safety'. IEE Digest 1997/295

Index

www.ingramcontent.com/pod-product-compliance
Lightning Source LLC
LaVergne TN
LVHW011754200926
848484LV00003B/704
9780852961070